Talk is Cheap

Talk is Cheap

The Promise of Regulatory Reform in North American Telecommunications

ROBERT W. CRANDALL
LEONARD WAVERMAN

The Brookings Institution / Washington, D.C.

1775 Massachusetts Avenue, N.W., Washington, D.C. 20036

Library of Congress Cataloging-in-Publication Data

Crandall, Robert W.
Talk is cheap : the promise of regulatory reform in North American telecommunications / Robert W. Crandall, Leonard Waverman.
p. cm.
Includes bibliographical references and index.
ISBN 0-8157-1608-7 (cl : alk. paper). — ISBN 0-8157-1607-9 (pa : alk. paper)
1. Telecommunication policy—United States. 2. Telecommunication—United States. 3. Telecommunication policy—Canada. 4. Telecommunication—Canada. I. Waverman, Leonard. II. Title.
HD7781.C66 1995
384'.041'097—dc20 95-37959
CIP

9 8 7 6 5 4 3 2 1

The paper used in this publication meets the minimum requirements of the American National Standard for Information Sciences—Permanence of Paper for Printed Library Materials, ANSI Z39.48-1984.

Set in Walbaum

Composition by Harlowe Typography, Inc.
Cottage City, Maryland

Printed by R. R. Donnelley and Sons Co.
Harrisonburg, Virginia

THE BROOKINGS INSTITUTION

The Brookings Institution is an independent organization devoted to nonpartisan research, education, and publication in economics, government, foreign policy, and the social sciences generally. Its principal purposes are to aid in the development of sound public policies and to promote public understanding of issues of national importance.

The Institution was founded on December 8, 1927, to merge the activities of the Institute for Government Research, founded in 1916, the Institute of Economics, founded in 1922, and the Robert Brookings Graduate School of Economics and Government, founded in 1924.

The Board of Trustees is responsible for the general administration of the Institution, while the immediate direction of the policies, program, and staff is vested in the President, assisted by an advisory committee of the officers and staff. The by-laws of the Institution state: "It is the function of the Trustees to make possible the conduct of scientific research, and publication, under the most favorable conditions, and to safeguard the independence of the research staff in the pursuit of their studies and in the publication of the results of such studies. It is not a part of their function to determine, control, or influence the conduct of particular investigations or the conclusions reached."

The President bears final responsibility for the decision to publish a manuscript as a Brookings book. In reaching his judgment on the competence, accuracy, and objectivity of each study, the President is advised by the director of the appropriate research program and weighs the views of a panel of expert outside readers who report to him in confidence on the quality of the work. Publication of a work signifies that it is deemed a competent treatment worthy of public consideration but does not imply endorsement of conclusions or recommendations.

The Institution maintains its position of neutrality on issues of public policy in order to safeguard the intellectual freedom of the staff. Hence interpretations or conclusions in Brookings publications should be understood to be solely those of the authors and should not be attributed to the Institution, to its trustees, officers, or other staff members, or to the organizations that support its research.

Foreword

THE LAST TWENTY-FIVE years revolutionized the telecommunications industries of most countries. Fiber-optics lines, high-speed digital electronic switches, local area computer networks, packet switching, and intelligent telephones have replaced the simple copper lines, rotary-dial phones, and electromechanical switches of the 1960s and 1970s. From simple voice telephone service, telecommunications is rapidly evolving into the broadband information superhighway of the future. These changes have placed strains on the old regulatory structure as new entrants have appeared to offer services that compete with the offerings of the old-line telephone companies and as the distinctions between telephony, computer services, and video services blur. Federal regulators in the United States have responded by opening up the U.S. telephone network to market entry, and Canadian regulators have generally followed a similar path, with a lag of about a decade. State regulators in the United States and provincial governments in Canada have generally tended to be more resistant.

In this book, Robert Crandall and Leonard Waverman trace the changes in federal and state or provincial regulation on both sides of the U.S.-Canadian border, finding that the form of regulation has indeed changed in response to the pressure of the times. Competition in long-distance services is increasing in both countries, but competition in the provision of local connections is only beginning. Unfortunately, neither country has succeeded in eliminating the distortions in regulated rate structures; thus, both are reluctant to allow full and open competition. The cost of maintaining these rate distortions may be as much as $30 billion in the United States and even more (on a per capita basis) in Canada. Crandall and Waverman conclude that more complete deregulation, except possibly for network interconnection rates, is required to eliminate these rate distortions and allow for the full exploitation of modern technology. Equally important, less regulation is crucial if meaningful competition is to develop in the expanding array of telecommunications services on both sides of the border.

Robert Crandall is a senior fellow in the Economic Studies program at Brookings, and Leonard Waverman is professor of economics and director of the Centre for International Studies at the University of Toronto.

The authors wish to thank Michael Deny, Melvyn Fuss, Charles Jackson, Paul M. MacAvoy, David E. M. Sappington, J. Gregory Sidak, Lester Taylor, William Taylor, and Clifford Winston for helpful suggestions and criticisms of early drafts of the manuscript. Research assistance was provided by Jeffrey McConnell, Jeffrey Santos, Nicholas Sisto, and Stephanie Wilshusen. The manuscript was edited by Steph Selice and Deborah Styles and checked for factual accuracy by Laura Amin and Cynthia Iglesias. Carlotta Ribar proofed the pages, Mary Mortensen created the index, and Lisa Guillory and Colette Solpietro provided administrative assistance.

Funding for this project was provided by the Center for Economic Progress and Employment, whose supporters comprise Donald S. Perkins; Aetna Life and Casualty Company; the American Express Philanthropic Program; AT&T Foundation; The Chase Manhattan Bank, N.A.; Cummings Engine Foundation; The Ford Foundation; Ford Motor Company Fund; General Electric Foundation; Hewlett-Packard Company; Institute for International Economic Studies; Morgan Stanley & Company, Inc.; Motorola Foundation; The Prudential Foundation; Alfred P. Sloan Foundation; Springs Industries, Inc.; Union Carbide Corporation; Alex C. Walker Foundation and Charitable Trust; Warner-Lambert Company; and Xerox Corporation. Additional funding specifically for this project was provided by the Alex C. Walker Foundation and Charitable Trust and Warner-Lambert Company.

The views expressed in this book are those of the authors and should not be ascribed to any of the persons or organizations acknowledged here, or to the trustees, officers, and other staff members of the Brookings Institution.

MICHAEL H. ARMACOST
President

Washington, D.C.
January 1996

Contents

Figures

Talk is Cheap

Chapter One

The Telephone Industry in the United States and Canada

THE NEW TELECOMMUNICATIONS infrastructure, sometimes dubbed the "information superhighway,"[1] will be built in part from the current facilities of the telephone industry. Indeed, the telephone industry is already gearing up for the change, replacing copper wires with fiber-optic cables and preparing for the delivery of video services over a network that was designed for much more modest voice telephony.

For most of its history the telephone industry has been organized as a monopoly—in most countries as an adjunct to the post office. In the past two decades, however, there have been significant changes in the ownership and competitive position of the old, established telephone carriers in many countries, including the United Kingdom, New Zealand, Japan, and Australia.[2] In many ways the United States has led this movement toward a more competitive and less regulated telephone sector, and Canada has lagged about a decade behind. In the past few years, however, Canada has struggled mightily to catch up with and even pass the United States in this regard.[3] Both countries have had a significant advantage in making this transition because the United States had not allowed the government to own and operate its telephone network, and

1. This term was apparently coined by Vice President Al Gore.

2. New Zealand and the United Kingdom have both privatized their national telecommunications carrier and opened their markets to competition. Australia has liberalized entry into its telecom markets, and Japan has partially privatized Nippon Telegraph and Telephone Corporation (NTT) and allowed limited competition in its markets.

3. See chapter 2 for the full details.

Canada had only limited public ownership.[4] The European countries, with the exception of the United Kingdom, must wean their telephone operations from public ownership. Several of these countries, most notably France, Germany, and the Netherlands, have finally begun this process.[5]

Nevertheless, in most countries, the telephone industry is still heavily regulated. Canada and the United States are no exception. Regulation exists in part to control potential problems of monopoly pricing, but regulation also *impedes* competition. Regulation is also responsible for severe distortions in pricing that, in turn, become the rationale for not allowing an immediate transition to competition. This book details the extent of these distortions, the effects of regulatory reform on both sides of the U.S.-Canadian border, and the effectiveness of competition in moving rates toward more economically rational levels. It begins with an overview of the telephone industries on both sides of the U.S.-Canadian border and a review of the changing national policy toward telephone competition and regulation, then describes the changes in industry structure, the evolution of new services, and the major players in both countries.

The End of Monopoly

Until 1960 the Canadian and American monopoly providers of telephone service were protected from competitive entry by government regulatory authorities. AT&T, Bell Canada, and the other major telephone companies were vertically integrated in the provision of local and long-distance service in their respective service areas. AT&T owned local telephone companies that had franchise monopolies in communities accounting for more than 80 percent of U.S. telephone subscribers and provided virtually 100 percent of interstate long-distance service. Bell Canada has had a similar po-

4. The provincial governments in the three prairie provinces owned and operated their telephone systems, but all but one of these systems have now been privatized. The largest telephone companies—Bell Canada and British Columbia Telephone—were always private.

5. In addition, the European Commission is applying strong pressure to force member states to allow entry into a variety of services now and into all services by 1998.

sition in the delivery of local telephone service in Ontario and Quebec, which together account for more than 60 percent of Canada's telephone subscribers. Today competition in long-distance (and cellular) calling exists in both countries, competitive local-access providers are growing (particularly in the United States), and cable television and telephone companies are girding for a competitive battle to provide the major links in the information superhighway. Despite the growth of competition in this industry, most telephone services are still regulated under a theory that they are supplied by monopolists or dominant providers, and Chinese walls are placed around potential competitors.

U.S. Regulation and Antitrust

AT&T's monopoly position came under attack in the United States more than thirty years ago. In 1959 the Federal Communications Commission (FCC) first approved the private use of microwave by large businesses desiring to establish their own networks. In 1969 the FCC extended this privilege to a new communications carrier, Microwave Communications, Inc. (later MCI), which was allowed to offer dedicated private-line circuits to business subscribers who could not or would not build their own networks.[6] MCI grew as a private carrier in the early 1970s by undercutting AT&T's rates and offering digital data services that AT&T at first did not offer.[7]

In the mid-1970s the assault on AT&T's monopoly intensified. MCI began to offer ordinary switched long-distance service without FCC approval;[8] when the FCC attempted to block it from continuing

6. Private lines are dedicated circuits leased to business customers. These circuits allow two-way communication between fixed points; they do not allow the lessee to access the switched network and thereby gain access to other telephone subscribers.

7. The history of the development of competition in the U.S. telephone industry is recounted in Gerald W. Brock, *The Telecommunications Industry: The Dynamics of Market Structure* (Harvard University Press, 1981). Another view may be found in Peter Temin (with Louis Galambos), *The Fall of the Bell System: A Study in Prices and Politics* (Cambridge University Press, 1987). See also Robert W. Crandall, *After the Breakup: U.S. Telecommunications in a More Competitive Era* (Brookings, 1991).

8. Switched long-distance service is often called message-telephone service (MTS). It is a toll service that allows the caller to access any other subscriber on the network through the switching machines of the long-distance company and the relevant local exchange companies.

to offer this service, MCI prevailed in a court ruling. At the same time, over the objection of state regulators,[9] the FCC forced telephone companies to allow customers to connect competitive equipment to their lines, ending the monopoly over terminal equipment—handsets, answering machines, modems, private branch exchanges (PBXs), and key telephone sets—held by AT&T (and other telephone companies).

The development of competition in terminal equipment and long-distance services created substantial controversy over the terms by which competitors could gain access to AT&T's local circuits. This access was crucial to the competitors because the local telephone companies controlled the only path to most telephone subscribers. Understandably, AT&T used the regulatory process and various dilatory tactics to slow its competitors' growth, arousing the suspicions of the antitrust authorities. In late 1974 the Department of Justice filed a Sherman Act antitrust suit, alleging that AT&T had attempted to use its local-access monopoly to monopolize equipment and long-distance markets.[10] This suit was settled in 1982 with a consent decree that required divestiture of the Bell local operating companies (BOCs) from AT&T into seven regional companies. The divestiture was carried out on January 1, 1984. The divested BOCs were allowed to offer local service and limited nearby long-distance services. AT&T was left with the bulk of the long-distance services, the manufacturing operations, and its research arm, Bell Laboratories.[11]

Under the terms of the 1982 AT&T decree, the seven regional Bell operating companies (RBOCs) were forbidden to engage in manufacturing or in providing long-distance services outside each of their local access and transport areas (LATAs).[12] In addition the

9. The jurisdictional struggles between federal and state authorities are detailed in chapter 2.

10. An earlier (1949) suit had been settled by a consent decree in 1956. It was this decree that was modified in 1982 to become the modified final judgment (MFJ) by consent agreement among the parties. *United States* v. *American Telephone & Telegraph Co.,* 552 F. Supp. 131 (D.D.C. 1982), aff'd sub nom., Maryland, 460 U.S. 1001, 103 S. Ct. 1240, 75 L. Ed. 2d 472 (1983).

11. A new research organization, Bellcore, was established for the RBOCs, staffed in part by former employees of Bell Laboratories.

12. There are 161 of these LATAs. Thirty-eight states have more than one LATA; twelve are single-LATA states. The local BOC may offer service within its LATA, but not between LATAs, even those within a state. The RBOCs were also forbidden to

BOCs were ordered to convert their switching networks to provide equal access to competitive long-distance providers. As a result, competition began to develop much more rapidly in the inter-LATA toll market, both within the multi-LATA states and between states. Long-distance service is now provided by three major carriers, several other facilities-based carriers,[13] and a host of resellers.

The 1982 decree created a balkanized U.S. telephone industry. The divested RBOCs were confined to local service and intra-LATA long-distance services. AT&T was left as principally a long-distance service provider and manufacturer of communications equipment. In 1984, Congress enacted a statute that dramatically reduced the regulation of cable television and banned telephone companies from providing cable television services in their local franchise areas.[14] Thus the major U.S. local telephone companies were excluded from long-distance services, cable television services, equipment manufacturing, and even information services in the mid-1980s.[15]

Twelve years have now passed since the AT&T breakup and the 1984 Cable Act. A revolution in electronics and digital compression technology is reducing the cost of providing a wide array of telecommunications services over copper wires, coaxial cables, fiber-optics cables, and radio circuits. The distinctions among services are beginning to blur as are the distinctions among the participants. AT&T, for example, has now reentered local telephone markets through its purchase of one of the largest U.S. cellular telephone companies, McCaw Cellular. MCI, the second largest U.S. long-distance company, is preparing to spend $2 billion to enter the local-access market, and Sprint has joined with three cable televi-

---

offer information services, but this prohibition was subsequently reversed by an appellate court. *United States* v. *American Telephone & Telegraph Co.,* 552 F. Supp. 131 (D.D.C. 1982), aff'd sub nom., *Maryland* v. *United States,* 460 U.S. 1001, 103 S.Ct. 1240, 75 L. Ed. 2d 472 (1983).

13. Facilities-based carriers are those that own all or part of their distribution network; resellers are those that exist largely as arbitragers, reselling the services of other carriers to final customers.

14. Cable Communications Policy Act of 1984, P. L. No. 98-549, 98 Stat. 2779 (1984). However, the FCC had already imposed cross-ownership restrictions on cable telephone companies before the 1984 legislation.

15. The ban on information services was eventually dropped as a result of Bell company challenges to the AT&T court's attempt to maintain the restrictions in enforcing the 1982 decree. See footnote 12.

sion companies to bid successfully for personal communications services (PCS) licenses in twenty-nine of the country's fifty-one major trading areas.[16]

The RBOCs are also extending their reach by purchasing interests in cable television companies outside their telephone franchise regions. These cable acquisitions and joint ventures appear to be motivated by the telephone companies' desire to use the cable companies' broadband networks[17] outside their own telephone franchise areas to build sophisticated, interactive networks capable of delivering video, data, and voice services. Thus competition among large telephone companies may develop in a manner quite unexpected by the parties who initiated the consent decree that broke up AT&T. In Canada, meanwhile, the major telephone carriers have decided to take another path by forming a large joint venture—presently called the Beacon Initiative—to build a national Canadian broadband information superhighway to compete with the country's cable systems.

At the same time the RBOCs are petitioning the AT&T trial court for relief from the decree's restrictions on their participation in manufacturing and inter-LATA markets.[18] Several have even contested the constitutionality of the 1984 Cable Act's restrictions on telephone company provision of cable television services. These actions have generally been successful, although the issue has not yet been decided by the U.S. Supreme Court.[19] And legislation is now pending before the U.S. Congress that would repeal the cable-telephone company ban and the line-of-business restrictions required by the modified final judgment (MFJ). Thus court appeals or legislation may soon reverse the vertical fragmentation occasioned by the 1982 *AT&T* decree as telecommunications firms seek

16. PCS is a new cellular-like service that is likely to compete with cellular and local wire-based terrestrial networks to offer local service. See the discussion of PCS below and John J. Keller, "MCI Says It May Have to Work Alone in Pursuing National Wireless Network," *Wall Street Journal*, October 24, 1994, p. A3.

17. "Broadband" refers to the ability of a communications network to transmit large amounts of information per unit time. Such capacity is necessary, for instance, to provide video, or television, signals.

18. *U.S.* v. *Western Electric, Inc. and American Telephone and Telegraph Company*, Civil Action No. 82-0192, U.S. District Court for the District of Columbia, *Motion of Bell Atlantic Corporation, Bell South Corporation, NYNEX Corporation, and Southwestern Bell Corporation to Vacate the Decree*, July 6, 1994.

19. See the discussion in chapter 8.

greater integration and the ability to compete with one another in the delivery of ever more complex services.[20]

Attention is now shifting to the states, which have traditionally been less receptive to proposals for liberalizing entry. Most states still do not allow competition for local residential telephone service, and some still prohibit full competition in the shorter long-distance bands. Over time, however, this resistance is being overcome and states are even beginning to experiment with less rigid forms of regulation.

The Canadian Approach

In Canada the path to liberalization was more deliberate and the role of the antitrust authority (which concentrated on the effects of vertical integration between Bell Canada and its manufacturing subsidiary) was much more muted than in the United States. In 1976 the director of the Bureau of Competition Policy petitioned the Restrictive Trade Practices Commission (RTPC) to require Bell Canada to divest itself of Northern Telecom, its manufacturing subsidiary. In 1983 the RTPC demurred, concluding that the various portions of the telecommunications equipment market were competitive and that vertical integration had real economic benefits.[21] Regulators in Canada have been concerned with the level and structure of prices but not directly concerned with vertical integration. As a result there has been no antitrust or regulatory examination of the potential for breaking up Bell Canada or B.C. Telephone in the manner that U.S. antitrust authorities obtained divestiture of AT&T's operating companies.

In 1979 the Canadian Radio and Telecommunications Commission (CRTC) authorized competition for nonswitched data services,

20. The authors have filed affidavits on behalf of the RBOCs in several proceedings. Crandall's discusses the virtues of RBOCs' offering of various services in areas outside the territories of their telephone operating companies. Waverman's discusses the vertical integration between Bell Canada and its manufacturing subsidiary, Northern Telecom, and the antitrust and regulatory examinations in Canada of this vertical integration.

21. "Bell had lower prices in every equipment category and, overall, had a purchasing bill on Northern's equipment lower by at least six percent in the three years; the average was just over seven percent." Restrictive Trade Practices Commission, Consumer and Corporate Affairs, Canada, *Telecommunications in Canada* (Ottawa 1983), part 3, "The Impact of Vertical Integration on the Equipment Industry," p. 110.

but this decision was limited to the provinces of Ontario, Quebec, and British Columbia because the CRTC's authority did not appear to extend to telephone services in the prairie or maritime provinces. In 1982 the CRTC liberalized terminal equipment attachment rules in the same geographic region. In 1984 competition in services was liberalized to a minor extent when the CRTC allowed very limited resale and shared use of private-line services. It was not until 1990 that resale and sharing of ordinary switched long-distance services was permitted.

In 1989 the Canadian Supreme Court ruled that all interconnected telephone companies, except provincially owned companies, were subject to federal jurisdiction and, therefore, the CRTC's regulatory authority.[22] It was not until 1992, however, more than twenty years after MCI's entry into switched-message toll service (MTS) in the United States, that the CRTC moved to allow full competition in long-distance services. After first denying CNCP[23] entry into basic switched long-distance services in 1985, the CRTC reversed course in 1992 and admitted Unitel (formerly CNCP) into the long-distance market. The Canadian Parliament passed the Telecommunications Act of 1993, extending the CRTC's authority to all intraprovincial and interprovincial telecommunications in Canada except in Saskatchewan.[24] Thus the CRTC's liberalization of long-distance and terminal-equipment markets is now in effect in most of Canada. In 1994 this liberalization was extended even to local services when the CRTC, in a far-reaching decision, allowed entry into local telephony. The new regulation required the incum-

22. *IBEW* v. *Alberta Government Telephones*, Supreme Court of Canada, August 14, 1989. For further discussion of the evolution of Canadian regulatory policy, see chapter 2.

23. A joint venture between a Crown corporation, the Canadian National Railway, and the Canadian Pacific group of companies.

24. Saskatchewan Government Telephone will be regulated by the CRTC in 1998. There are a number of small telephone companies in Canada ranging from rural coops having several hundred members to Thunder Bay, a municipally owned system of 78,700 main lines, and Quebec Telephone Company, a system in Quebec with 240,000 subscribers. A question remaining from the 1989 Supreme Court decision, whether these telephone companies, co-ops, and municipal franchises were under CRTC jurisdiction, was resolved on April 26, 1994, when the Supreme Court of Canada issued a decision that these companies fall under federal (CRTC) jurisdiction, because these companies provide interprovincial services. Judgment, Supreme Court of Canada, April 26, 1994.

bent local carriers to unbundle their networks and permitted them to offer video dial tone in competition with cable companies.[25] Thus in just two years Canada appeared to have leapfrogged the United States, which retains a large number of state-imposed barriers to local competition. Unfortunately, the Canadian government delayed the implementation of the CRTC's 1994 decision for one year (see footnote 61, chapter 2, p. 68).

With liberalization in Canada has come a substantial amount of integration between U.S. and Canadian carriers. AT&T purchased a 20 percent share of Unitel in 1993; Bell Canada has purchased a 5 percent interest in MCI; and Sprint has acquired an interest in Call-Net, Canada's third largest long-distance company, renaming it Sprint Canada. The largest reseller in Canada, ACC Tel Enterprises, is a subsidiary of ACC Corp. of Rochester, and Bell Canada Enterprises (BCE) has purchased an interest in the U.S. cable company Jones Intercable.

Further Canadian-U.S. integration is likely in the wake of the Canada–U.S. Free Trade Agreement and the North American Free Trade Agreement (NAFTA). In addition, high Canadian interprovincial long-distance rates induce large customers to bypass the Canadian network through U.S. routes south of the border.[26] Thus market liberalization in one country inevitably creates pressure for similar changes in the other.

The Fragmenting Telecommunications Network

As a result of these past two decades of liberalization and the recent wave of mergers and joint ventures, the once monolithic telephone network in the United States has been fragmented and reconnected in new ways. Increased competition in Canada is likely to have a similar effect on the vertically integrated telephone companies. Customers, new competitors, and traditional telephone companies are interconnecting in new ways. The rapid changes in electronics and communications technology interact with these structural changes to create a confusing array of choices for customers and a vastly more difficult problem for regulators. The

25. CRTC Telecom Decision 94-19, September 16, 1994.

26. See Table 6-8 for a comparison of U.S. and Canadian interstate (or interprovincial) rates.

number of actual and potential competitors at each level or stage of the industry is increasing rapidly. Many of these competitors require interconnection with downstream or upstream firms that also compete with them. Thus they are quick to complain to regulators that these connections are not available at fair prices or that they face unfair competition subsidized by revenues flowing from regulated services. The result is a tangled web of regulated and unregulated firms rapidly expanding into new markets through new technologies. The telephone industry is not what it used to be.

This fragmentation is best understood by focusing on each of the major services provided by the telecommunications system. Among the most important of these services is access, or the connection of users to the network. In residential services, access simply comprises the handset and the pair of copper wires that presently connect it to the telephone company's switch. For a business, access may be attained via a modem or a much more complicated device that delivers hundreds of calls or large amounts of data through a variety of lines to the telephone company's or a competitor's switch or even directly to another location. Increasingly the equipment on the customer's premises is gaining in intelligence and is itself capable of switching calls. "Lines between equipment and transmission services are becoming equally blurred. Such things as very small aperture terminals (VSATs) and satellites are all terminal equipment in that they are discrete units located at one end or another of a transmission path . . . they do not merely use the connection, they create it."[27]

Access has no value, however, unless the access line is used. Use may involve the delivery or reception of voice, data, or video signals, each of which may be priced quite differently by telephone companies. Traditionally, residential voice lines and even single business lines have been charged a flat monthly rate for unlimited use over a defined local calling area, but long-distance services and more sophisticated services have generally been metered and a charge levied for their use, often at rates that vary by time of day.[28]

27. Peter W. Huber, Michael K. Kellogg, and John Thorne, *The Geodesic Network II: 1993 Report on Competition in the Telephone Industry* (Washington: The Geodesic Company, 1992), p. 6.45.

28. Long-distance services were traditionally referred to as "toll" services and

Local Access

For more than a century, access to the local network was controlled by local telephone companies who owned local exchanges, switching centers that were connected to all subscribers by paired copper wires. Small companies might use only one telephone switch to connect all of their subscribers, but as the size of the local access-exchange area increases, the required number of switches increases, and the company must interconnect the various switching centers. These terrestrial systems require expensive rights-of-way and the commitment of extensive sunk capital in telephone poles, conduits, wires, switches, and other equipment.

Other technologies for providing telephone access have promise. Wireless access may be provided by mobile radio; in recent years, this mobile service has been provided through a cellular architecture in order to use radio frequencies more intensively. Digital radio, connecting residences and businesses without the use of wires, may also be used as a substitute for the fixed terrestrial network. Alternatively, access may be provided to large customers through dedicated microwave circuits or large-capacity fiber-optics networks that are currently economical only in densely populated business districts.[29] Low-orbiting satellites also promise to provide wireless access over vast distances.

The large fixed, sunk costs of extending wires from central-office switches to an individual subscriber's premises may preclude competitive entry by firms using the same technology as the traditional telephone companies. However, an adequate market test of contestability in traditional telephony has not occurred because entry has traditionally been blocked by regulators. Entry into local access-exchange service by radio-based technologies is now developing rapidly, and entry by wire-based facilities (such as cable television) may be imminent.

The first major expansion of wireless service involved the development of cellular telephony, a service that allows owners of

---

were different from local calls in that they required substantial additional costs of switching and transmission. As switching and transmission costs have decreased, however, the distinction between a local call and a toll call has become essentially arbitrary.

29. See chapter 7 for a discussion of the prospects for competition in local access.

portable telephone receivers to communicate through the spectrum at relatively low power with transmitters spaced throughout a geographic region in discrete cells. This architecture allows cellular providers to reuse the same frequency several times over a given metropolitan area and thus to economize on scarce spectrum.

Cellular service developed somewhat differently in the United States and Canada. In the 1970s the U.S. Federal Communications Commission (FCC) decided to award two cellular franchises in each metropolitan market and, subsequently, in rural service areas, with one franchise in each market reserved for the local wire-line telephone company. After long licensing delays, U.S. cellular service finally commenced in 1983 in Chicago. Because of the FCC's decision to license cellular on a market-by-market basis, no truly national cellular service has developed in the United States.[30] The CRTC followed the FCC's model of licensing two cellular carriers, but it asked for bids for two *national* licenses. Cellular service began in Canada in 1988 with the provincial telephone companies holding one cellular license, and a national competitor, Cantel, holding the other. Cantel's majority owner is Rogers, the holder of 29.5 percent of the voting stock in Unitel and the largest operator of cable television in Canada.[31]

The possibilities of using radio-based communications are expanding steadily as regulators open up more electromagnetic spectrum for such uses and as digital compression techniques allow more and more information to be transmitted over a limited bandwidth. These new radio-based technologies are generically referred to as "personal communications services" (PCS), reflecting the ability of the individual to carry his own personalized transceiver and telephone number rather than having to be in his home, place of business, or vehicle to send and receive messages. Some of these new PCS networks may be built as complements to cable networks or long-distance networks, combining wire or satellite

30. A number of companies have pursued an aggressive strategy of acquiring cellular franchises, but none has more than 14 percent of U.S. cellular subscribers. The largest nine companies have almost 80 percent of subscribers. Dennis Liebowitz and others, *The Wireless Communications Industry*, Donaldson, Lufkin, and Jenrette Securities Corporation, Winter 1994–95, p. 11.

31. Harvey Enchin, "Rogers Writes Off Unitel," *Globe and Mail*, August 10, 1995, pp. B1, B4.

interconnection of distant users with local radio-based connections.[32] As this book is being written, the United States is in the process of auctioning spectrum for PCS use. The first two blocks of this spectrum brought $7.7 billion in bids in 1995.[33]

In the United States and Canada, cable television companies currently serve more than 90 percent of all residences with a "bus" network, consisting of coaxial cable and, to a lesser extent, fiber optics for feeder lines. These systems are designed to deliver a common set of signals in one direction to every subscriber (see figure 1-1). Switched telephone networks, meanwhile, use some variant of a star network to provide unique two-way communications to every subscriber. Each subscriber must be connected to a switch that can route his outgoing or incoming call to its destination (figure 1-1). These connections are generally accomplished with pairs of copper wires whose capacity is limited to voice and low-speed data applications. Cable networks have much greater capacity because they deliver scores of video signals over coaxial-cable drop lines, but with their current network architecture, they cannot provide switched two-way communications. These cable networks could be modified to deliver high-capacity *switched* communications services if the cable systems were to redesign the network architecture and to install switching machines. A number of such experiments are currently under way.[34]

An alternative approach, now in use in the United Kingdom, is to wire the cable network with both coaxial cable and paired copper-wire drop lines, connecting the latter to traditional telephone switches for the delivery of telephone service. While this architecture may be retrofitted into existing coaxial-cable networks in Canada or the United States, such an exercise is much more expensive than wiring both facilities when the cable system

32. Motorola's IRIDIUM Project and the announced joint venture between Bill Gates and Craig McCaw that will use a large number of orbiting satellites to provide ubiquitous mobile service are two examples of emerging plans to bypass the local fixed network.

33. The auction began in December 1994 and was completed in March 1995. See chapter 7 and Dan O'Shea, "PCS Auction Survivors Await Next Battle," *Telephony*, March 20, 1995, pp. 6–7.

34. Time Warner is currently building such a system into its Rochester, New York, cable television operations. TCI, the largest cable system in the United States, has also announced that it will spend several billion dollars to build a switching capacity into its cable systems. See chapter 7.

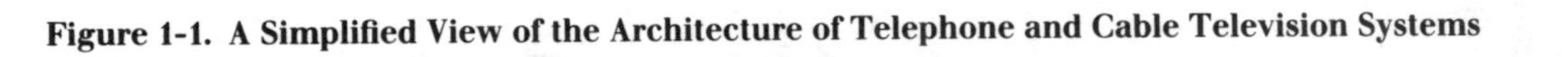

Figure 1-1. A Simplified View of the Architecture of Telephone and Cable Television Systems

Telephone "Star" Network

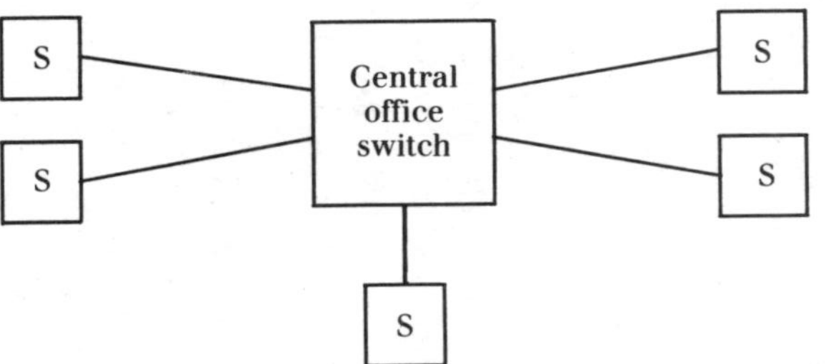

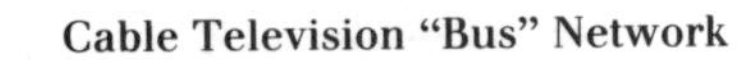

Cable Television "Bus" Network

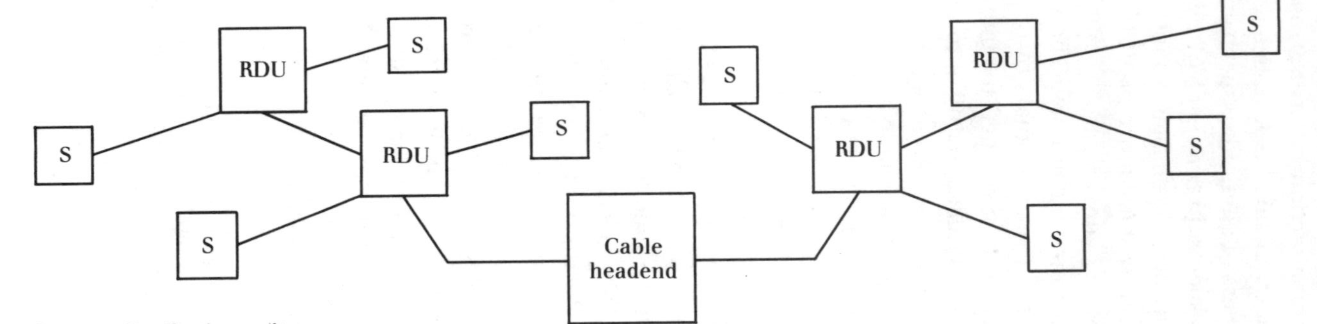

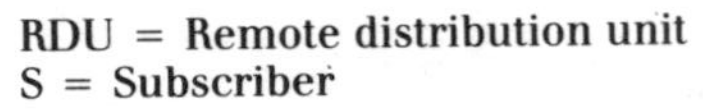

RDU = Remote distribution unit
S = Subscriber

is originally built. Thus direct cable television competition for telephony services in Canada and the United States is less likely than in the United Kingdom.

These revolutionary changes in telecommunications technologies have not yet created a truly contestable market for local access, partly because of regulatory policy. Nevertheless, the notion that local-access services are subject to large economies of scale and are thus a natural monopoly is clearly being challenged. Cellular services, a second wire-line service offered through the cable-television infrastructure that is already available to most homes, and a variety of other new wireless services may eventually end the traditional telephone company's control of local services. Regulators have been loath to allow competition for the local loop in many jurisdictions or have so distorted rates as to make entry into local markets unattractive. In Canada, for example, Bell Canada incurred losses in 1989 equal to almost $2 billion (current Canadian dollars) in its local access-usage services, which were financed by a similar amount of excess profits from its long distance services.[35]

Long-Distance Services

The distinction between local and long-distance services is necessarily arbitrary. Connecting a subscriber by wire, radio, coaxial cable, or fiber optics to the nearest telephone switch is clearly an *access* service. Thereafter, the call is switched and transmitted to its final destination. The costs of such transmission and switching depend on call density, distance, and time of day, but there is no obvious dividing line between local and long-distance service. In a dense market a call may be transmitted between switching centers as little as two or three miles apart. In other markets a call between

35. These data reflect the implementation of cost accounting conventions prescribed by the CRTC (Phase III rules for 1986–89). They are not necessarily an accurate reflection of economic costs, but they provide at least a rough indication of the underpricing of local service in the amount of $250 per line per year. Melvyn Fuss and Leonard Waverman, "Efficiency Principles for Telecommunications Pricing: Fairness for All," in Steven Globerman, W. T. Stanbury, and Thomas A. Wilson, eds., *The Future of Telecommunications Policy in Canada* (Bureau of Applied Research, Faculty of Commerce and Business Administration, University of British Columbia, and the Institute for Policy Analysis, University of Toronto, April 1995).

persons twenty-five or thirty miles apart may be handled by a single switching center.

In the United States and Canada, the definition of long-distance service often depends on regulatory practices. Small local, flat-rate calling areas, often supplemented by message units for longer calls in the same metropolitan area, are preferred by state regulators in the United States. In Canada, local flat-rate service is generally available over much wider calling areas through extended-area service (EAS). Within an EAS all calls are local. Given the zero price of these local calls, the larger the EAS, the less feasible is competition in long-distance services because entrants will find it more difficult to aggregate sufficient traffic to operate profitably.

In both countries, long-distance service is offered within and between provinces or states and, in the United States, often under different regulatory constraints because of the fifty-one separate regulatory regimes. Service is offered under a variety of pricing plans in both countries, allowing subscribers to choose from an array of optional tariffs. In addition, bulk discount plans are available to those with large volumes of long-distance traffic.

Given the differences in telephone rates among customers of different sizes, the opportunities for arbitrage are substantial. As a result, resellers have become an important competitive element both in the United States and, since 1990, in Canada. These carriers, who typically own very little switching and transmission equipment, offer to assemble low-cost network services by choosing among the bulk offerings of several large and small carriers and reselling these services to a variety of smaller customers, thus providing market arbitrage. These resellers compete with the facilities-based carriers, the larger national firms such as AT&T, Bell Canada, Unitel, MCI, or Sprint, and the smaller regional carriers such as Alascom.

An important constraint on the development of competition in long-distance or local services is the degree to which scale economies exist in the production of these services. In the 1970s new long-distance competitors in the United States used a mixture of microwave facilities and leased AT&T circuits to assemble a network. These microwave facilities were not subject to large-scale economies; therefore, it appeared that long distance was not a natural monopoly.[36] In the 1980s the development of fiber optics rev-

36. See Leonard Waverman, "The Regulation of Intercity Telecommunications,"

olutionized long-distance transmission, and every national carrier replaced most of its microwave plant with extremely high-capacity fiber optics.

Given the scale economies in fiber-optics transmission, large economies of scale are now likely to exist once again in long-distance services. As a result, some students of the industry now question whether competition is feasible in this market.[37] However, long-distance services may also be provided by cellular carriers with or without terrestrial interconnection. Cellular systems could be interconnected by satellites, thereby bypassing the terrestrial network altogether (as in Motorola's IRIDIUM Project). In the United States, restrictions in the decree that divested the Bell Operating Companies from AT&T prevent the Bell-owned cellular companies from delivering a national cellular long-distance service.[38] The independent cellular carriers have no such limitations on their ability to offer national service, and, as noted, Cantel is the Canadian national cellular provider.[39]

Central-Office Services

Given the sophistication of current stored-program-control digital switches, local telephone companies are now able to provide their customers with much more than just switched voice and data services. Residential customers may now buy an array of central-office services, such as call waiting, message storage and retrieval, call forwarding, and automatic number identification. Business customers can obtain a variety of signaling, billing, and network control functions, often combined into software-defined networks. These custom services are generally less tightly regulated than are basic services.[40]

---

in Almarin Phillips, ed., *Promoting Competition in Regulated Markets* (Brookings, 1975), pp. 201–40.

37. See, for example, Huber and others, *The Geodesic Network II*, chapter 3.

38. This decree is discussed in more detail in chapter 2 and in Crandall, *After the Breakup*.

39. When AT&T acquired McCaw Cellular, however, it agreed to offer equal access to its long-distance providers in order to gain antitrust approval for the acquisition.

40. Fuss and Waverman question the unregulated prices set for these optional residential services, showing that they may be higher than socially desirable. Fuss and Waverman, "Efficiency Principles for Telecommunications Pricing."

At the same time as these new services were being developed by telephone companies, customers were investing in "smart" terminals that can switch and transform voice and data. Thus intelligence was beginning to reside in both telephone central offices and customers' own premises. Just where this intelligence should reside to minimize the total social costs of telecommunications is an important issue, and the resolution of this issue is likely to be affected by regulation.

Information Services

Today's digital technology also permits telephone companies to offer a variety of content-based information services. Yellow Pages or other advertising directories may be stored in memory banks and accessed by customers in a fraction of a second through standard telephone lines. Bulletin-board services, such as those provided by France Telecom through Teletel, are offered by some telephone companies in North America. Telephone companies may now even offer news services, weather services, sports lines, or other content-based services in competition with services provided electronically by other companies (over the telephone network) or with print media.

Information services provided by local-exchange carriers have been slow to develop in the United States and Canada, in part because of government regulation but also because of start-up costs of providing consumer terminal equipment. In the United States the court administering the AT&T decree generally forbade BOC participation in this market, but an appellate court decision overturned this restriction. However, most telephone companies appear to be moving very slowly to develop this market opportunity. The major information services, such as America Online and CompuServe, have been developed without RBOC participation.

New Services and Technologies

The rapid changes in electronics and communications technologies are creating a wide range of new opportunities for the telephone industry. These new technologies are reducing the cost of transmission speed, switching, and network control.

Digital Services

The original telephone network transmitted analog voice signals at very low speeds. When users began to demand the capability to transmit data, telephone systems either shunted these services onto dedicated digital lines or required the use of modems to convert the digital bit streams into signals that could be transmitted over analog voice channels. The latter solution was unsatisfactory because of slow transmission rates over voice channels and the potential for error.

As more and more business users sought digital services, often at higher data speeds, for their ordinary telecommunications services, the telephone industry developed a new integrated services digital network (ISDN) capacity. This service has generally been offered at two speeds: basic rate—two 64 kilobytes per second (Kbs) channels plus 16 Kbs signaling—and primary rate—providing twelve times as much capacity, 1.544 megabytes per second (Mbs)—the equivalent of twenty-four 64 Kbs voice channels. These digital services can accommodate a mix of voice, data, and even (less than full-speed) video services over the same line if used in conjunction with the appropriate terminal equipment.

At first, customer adoption of ISDN was rather slow. The United States has a smaller share of its central offices equipped to deliver ISDN than do many developed countries, but the United States appears to have experienced the greatest demand for ISDN. This demand is now increasing rapidly—352,000 lines had been installed by the end of 1994 compared with 99,000 at the end of 1992.[41] By 1995 more than half of all U.S. Bell operating company access lines will be connected to switches that are capable of delivering ISDN. Most of these ISDN circuits are apparently basic-rate ISDN, the slower of the two speeds discussed above.[42]

Higher-Speed Circuits

ISDN is only the beginning of the technical revolution that is possible. Most developed countries are currently developing much

41. Federal Communications Commission, *Statistics of Communications Common Carriers,* 1992–93 edition, p. 19, and 1994–95 edition, p. 19.

42. Few data exist in Canada on the extent of ISDN development.

higher-speed switched services with as much as 150 Mbs speeds (equivalent to about 2,400 separate 64 Kbs voice channels on one path). Obviously only a few users could currently find a use for such communications rates, but some scientific and database applications could use rates much higher than those currently available.

All telephone companies are increasing their commitment to fiber optics in order to reduce their transmission costs. As fiber replaces copper transmission lines, transmission speeds are increasing from 1.5 Mbs (DS-1) to 45 Mbs (DS-3). Customers are now able to obtain 45 Mbs speeds by leasing DS-3 private lines for access to these networks. Thus large data customers are now able to obtain communications rates that are several thousand times as fast as those that were available over the voice network two or three decades ago.

Fiber-Optics and Video Services

The development of fiber optics and advanced digital compression techniques (which allow an expansion of the amount of information that can be squeezed through any transmission medium) has provided the telephone industry with a new potential market. By connecting their fiber-optics networks to customers through higher-speed access lines, telephone companies could offer households full video service—cable television. Many telephone companies are now planning to develop one-way and switched video services by using coaxial cables to connect their subscribers to the fiber-optics backbone of their networks.[43] Indeed, it is even possible that video service can be delivered through the current copper paired-wire access lines to households using an advanced digital compression technology.[44] A full-motion video signal requires a bandwidth of about 80 Mbs for a rapid-speed sporting event, the equivalent of about 1,250 voice channels. However, digital-signal compression techniques (such as MPEG-2) can reduce this to 4 to 6 Mbs.

43. See chapter 7 for a discussion of the costs of building such a fiber-coaxial cable network.

44. One of these technologies, asynchronous digital subscriber loop (ADSL), is now being tried on a pilot basis by a number of U.S. local exchange companies.

The prospect of telephone networks delivering video services is both novel and quite disruptive to the established regulatory order. Local telephone companies are universally regulated as monopolists of (wire-line) local access services. Entry by these regulated companies into the video market in competition with cable systems triggers concerns about the possibility of cross subsidies from regulated local access-exchange services to new video and information superhighway services. In the United States telephone companies are barred by the 1984 Cable Act from offering cable television service in their own franchised areas, but the FCC has ruled that they may offer video dial tone—the video circuit to the subscriber—although the commission has moved very slowly to approve specific telephone companies' proposals to offer this service.[45] The companies may not, however, control the programming decisions over these circuits.[46]

Similar regulatory constraints exist in Canada. Telephone companies are prohibited from offering cable television, but they may offer trials of video dial-tone service over their networks as long as they do not control the programming.[47] Cable companies in Canada are moving to become alternative access providers, offering unswitched private line access to bypass the local telephone company, although they argue that they are not presently contemplating competition in switched telephone services.

Three announcements show that the division between telephone on the one hand and video on the other may be about to break down. In March 1994 Rogers Cable, Canada's largest cable television company, acquired the third largest cable company, MacLean Hunter. Rogers, as noted earlier, is the majority owner of Cantel, the national cellular competitor to the telephone companies' cellular systems, and then owned 29.5 percent of Unitel, the

45. FCC, CC Docket 87-226, *Order* (1992).

46. This ban may soon be lifted by the Congress as part of a comprehensive reform of the nation's telecommunications regulatory apparatus. For instance, in the 104th Congress, the chair of the Senate Commerce Committee, Senator Larry Pressler (R-S.D.), offered a bill that included repeal of the ban on telephone-company provision of video programming services. Moreover, the 1984 ban has been ruled to be in conflict with the First Amendment. See *Chesapeake and Potomac Telephone Co. of Virginia* v. *United States,* 830 F. Supp. 909 (E.D. Va. 1993).

47. CRTC Telecom Decision 94-19. See also Public Notice CRTC 1994-130 undertaken in response to Order in Council P.C. 1994-1689.

principal long-distance competitor. On April 3, 1994, three days after Rogers announced its successful takeover of MacLean Hunter, the Stentor[48] group of Canadian carriers announced a $12 billion (Canadian) plan to upgrade the telephone network, providing fiber-optic or coaxial cable to all users. In late May 1994, a consortium of Canadian companies, including Bell Canada and Rogers Cable, announced a joint venture direct broadcast satellite (DBS) service to compete with Hughes' DirecTV, a U.S. DBS operation that began offering service in June 1994.[49]

At present, telephone companies are only exploring the possibility of delivering video services to subscribers, and, in some cases, underwriting small-scale experiments in video delivery.[50] With direct broadcast satellites able to offer as many as 150 channels of subscription television, new multichannel microwave distribution services (MMDS) able to offer up to thirty-three channels of service, and cable systems expanding their services from an average of forty-plus channels to potentially several hundred channels per system, entry into the video market is risky indeed.

The costs of entry into the video market for telephone carriers could be substantial. The current telephone company switched net-

48. Stentor is the corporation formed by Canada's established provincial telephone service providers to deliver interprovincial long-distance services. Stentor was established in 1992 to replace Telecom Canada, the earlier association of these telephone companies.

49. In late 1994 Power DirecTv, a joint venture between DirecTV and Power Corporation of Quebec, petitioned the government for permission to offer a version of the DirecTV service in Canada. The Canadian consortium (ExpressVu), now formally stripped of its cable partners except for the digitization and uplinking facilities, has received permission to operate under an exemption order. Power DirecTv asked for and was granted similar permission to offer a limited number of the U.S. DirecTV signals, the mandated Canadian signals, and 80 to 100 pay-per-view signals. Canada strictly limits the importation of U.S. signals that compete with Canadian programming, but it is powerless to prevent U.S. satellites from placing these signals over Canada. At present, an estimated 500,000 to 600,000 Canadians own satellite dishes to obtain these forbidden U.S. signals from U.S. satellites. DirecTV could greatly increase this number because its receiver-dish is likely to be much cheaper than conventional satellite dishes. Marian Stinson, "Another Rate Hike Looms," *Globe and Mail*, April 5, 1994, pp. B1, B6. The CRTC began public hearings on October 30, 1995, to consider applications for licenses for DBS services. There are three applicants: ExpressVu, Power DirectTv, and Homestar (CRTC News Release, August 31, 1995).

50. As this book goes to press, Pacific Bell, Bell Atlantic, and NYNEX appear to be shifting to a wireless video strategy as they experiment with various fiber-optics alternatives. In addition, Time Warner Enterprises (a joint venture with U S West) has begun to offer services in Orlando.

work is not equipped to offer wideband distributive services, but it could be modified to do so. Estimates of the costs of such modification are in the range of $1,500 per subscriber if fiber optics is required in the local subscriber loop, but substantially lower for a hybrid fiber-coaxial cable system (see chapter 7). Such an investment would essentially double the capital invested in the U.S. and Canadian telephone systems—to about $550 billion.

If telephone companies choose to develop a video service by using one of the new technologies for adapting their current paired copper-wire local loops, such as ADSL,[51] the required investment would be substantially lower for services with low subscriber penetration. However, this approach would allow only a few channels of service to enter the home at one time. The subscriber would have to order the desired service from a telephone central office, making "grazing" across channels difficult. On June 15, 1995, the Canadian government announced that three 30 MHz and three 10 MHz spectrum blocks would be licensed (not auctioned) in the 2 GHz frequency range. Existing cellular systems are restricted to 10 MHz PCS licenses, based on an announced maximum of 40 MHz for existing license holders.

Interactive Switched Video

The increasing capacity of fiber-optics systems and the possible expansion of telephone company subscriber lines through coaxial cable or ADSL has opened up the possibility of a variety of new switched broadband services, capable of delivering two-way video for telephony or video conferencing, video games, or information-intensive scientific or medical applications. The cable television companies are also exploring the possibilities of switched two-way broadband services, often in joint ventures with telephone companies. These prospective networks are at the center of policy discussions involving the development of the information superhighway and technological convergence. It is widely believed that all types of information—voice, data, and video—will be reduced to small digital packets and moved at the speed of light over the same

51. See footnote 44, p. 20.

fiber-optics network. Over time, as digital signal compression improves, the capacity of each of these networks will grow dramatically. There could be one, two, or more such networks, potentially interconnecting every home and commercial establishment. Whether such switched high-speed networks can provide services that are sufficiently valuable to consumers to merit their installation in every household remains an open question.

Personal Communications Services

The rapid evolution of microelectronics and digital communications has spawned an entirely new class of telephone services over the electromagnetic spectrum. Small hand-held transceivers may be used to communicate through remote antennas in buildings, on cable or telephone poles, or even on satellites. These devices may be designed to extend existing wire-line service or to provide a new stand-alone service with total number portability. Indeed, the convention of assigning telephone numbers to *locations* may be replaced by one that assigns numbers to *individuals* wherever they are. Telephone-company participation in these PCS markets may be limited by regulators' desire to increase competition for the local loop. Nevertheless, many U.S. and Canadian telephone companies are eagerly exploring this new technology, and U.S. firms are now ready to proceed after the FCC auction of two 30-MHz blocks in each major trading area.

Summary

Video service, ultra high-speed data services, and personal communications services are but three of the myriad of new services that may be offered by traditional telephone companies. Each market is likely to feature substantial competition, thus complicating the problem for telephone companies and regulators alike. As long as local exchange-access companies are tightly regulated in their basic service offerings, they will be constrained in their ability to enter competitive markets because regulators will be concerned that they will cross-subsidize these new ventures by overstating their costs in their regulated activities. Unfortunately, regulators will continue to press for these constraints as long as the telephone companies have a monopoly over the local loops that deliver telephone service. A vicious circle results: the existence of the local bottleneck monopoly perpetuates regulation, and regulators act to prevent entry. It is the regulators who are in large part responsible for perpetuating these local monopolies, often with the implicit encouragement of the local carriers, thus extending their own raison d'être.

The Current Structure of the U.S. and Canadian Telephone Industries

The telephone industries of Canada and the United States are part of a larger communications sector that accounts for about 3 percent of U.S. gross domestic product (GDP; see table 1-1). Included in the U.S. measure of GDP originating in the communications industry are radio and television broadcasting, cable television, satellite services, cellular companies, and telephone companies, most of which may soon be in competition with one another in a variety of voice-data-video markets. Note that there has been relatively slow growth in the contribution of communications to U.S. GDP since 1980 despite the extraordinary growth in new services. This slow growth may be largely the result of the sharp decline in the price of these services that is not captured fully in the price deflators, but it also results from the fact that a large share of communications is not measured: the official data exclude the services of privately owned networks, satellite dishes, local area (computer) networks (LANs), and telephone terminal equipment. For example, most large companies or facilities serve as their own telephone companies, routing voice calls through a private automated branch exchange (PABX) or a large key telephone system and data through their own LANs.

The Canadian data in table 1-1 include the GDP originating in the postal service, but it cannot be the postal service that was driving a stronger upward trend in this sector's share of GDP in the 1970s and 1980s. The telephone and cable television industries must have been responsible.

The slow growth of the public U.S. telecommunications network in recent years is also evident in the data on real capital stock in this sector (figure 1-2). After growing rapidly in the 1970s, the real

Table 1-1. GDP Originating in the Communications Industry as a Share of U.S. and Canadian GDP, 1970–93

	United States			*Canada*		
Year	*GDP originating from communications (billion 1987 $)*	*GDP (billion 1987 $)*	*Communications share of GDP*	*GDP from communications (billion 1986 $)*	*GDP (billion 1986 $)*	*Communications share of GDP*
1970	n.a.	2,874	n.a.	4.025	271.372	.015
1975	n.a.	3,222	n.a.	6.702	350.113	.019
1980	94.4	3,776	.026	10.320	424.537	.024
1985	115.8	4,280	.027	12.635	489.596	.026
1990	140.8	4,897	.029	18.044	565.155	.032
1991	148.2	4,868	.030	18.592	554.735	.034
1992	153.8	4,979	.031	19.015	558.165	.034
1993	158.9	5,134	.031	19.584	570.541	.034

Sources: U.S. Department of Commerce, Bureau of Economic Analysis; Statistics Canada.
n.a. Not available.

Figure 1-2. Real Gross Capital Stock

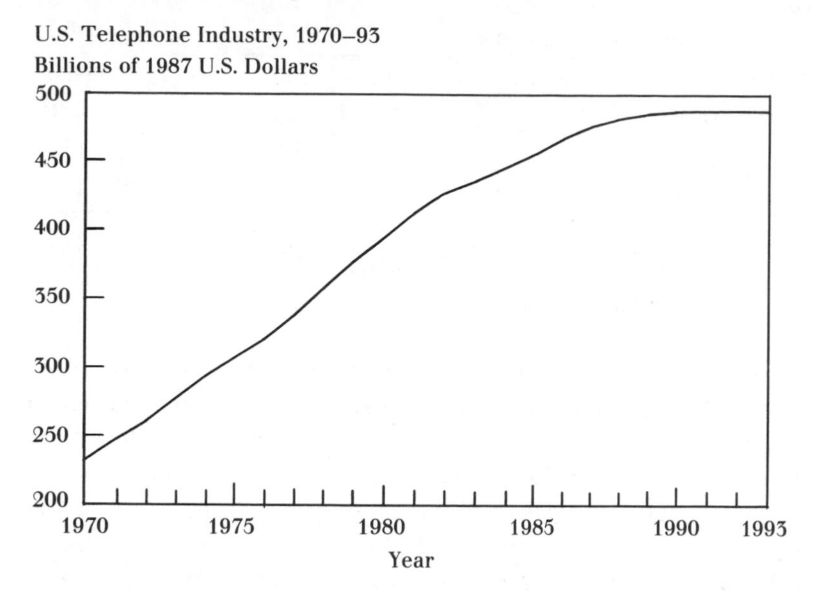

Source: U.S. Department of Commerce, Bureau of Economic Analysis, *Telephone and Telegraph*, July 8, 1994 (SIC 481, 482, 489).

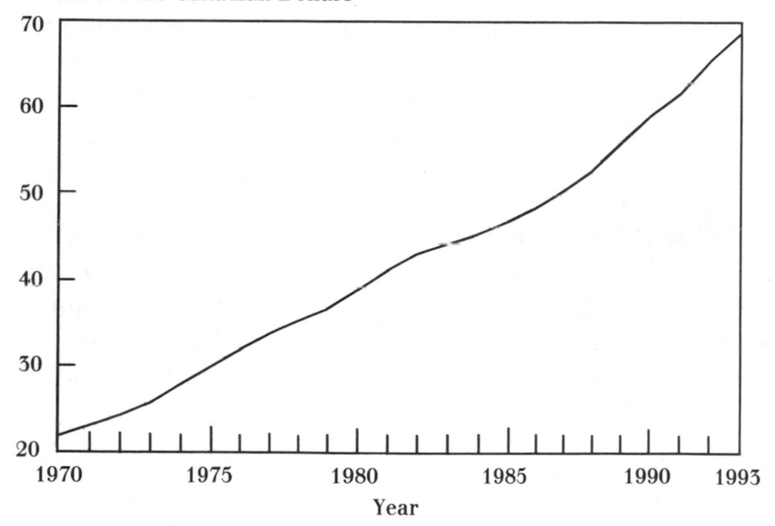

Source: Statistics Canada, CANSIM Database Matrix, no. 18-568, 1995.

Table 1-2. Sources of Revenues, U.S. and Canadian Telephone Industries, 1992

Billion U.S.$; percent in parentheses

Revenue source	*U.S.*		*Canada*	
Local services	48.1[a]	(37.4)	4.8	(42.5)
Long-distance services	73.0	(56.8)	5.8	(51.3)
Other services[b]	7.4	(5.8)	0.7	(6.1)
Total	128.5	(100.0)	11.3	(100.0)

Sources: United States: USTA, *Statistics of the Local Exchange Carriers*, 1992; FCC, *Statistics of Communications Common Carriers*, 1992/93. Canada: Statistics Canada, *Telephone Statistics*, Catalogue 56-203 (Canadian revenues converted to U.S. dollars at average exchange rate of 1992, $1.2081.)

a. Includes subscriber line charge.

b. Excludes nontelecommunications revenues.

stock of capital in the U.S. telephone industry has grown much more slowly since the mid-1980s despite the dramatic changes in technology and potential new services that it could offer. At the same time growth in the capital stock of the Canadian industry has actually accelerated. Much of the difference may be due to the line of business restrictions in the *AT&T* decree, but it may also be due to the growth in the United States of competitive media, such as private networks, satellites, and cable television.[52] Whatever the reason, the Canadian telephone industry has clearly been investing more aggressively than its U.S. counterpart.

The telephone industry is still largely a set of regulated firms providing switched voice services (plain old telephone service, or POTS) through a combination of fiber optics and paired copper wires. Its revenues derive almost entirely from local access-exchange services and long-distance services. In the United States 37 percent of revenues derive from charges for local access and use, 57 percent from long-distance services, and 6 percent from all of the other sources of revenues (see table 1-2). Telephone companies are still far from the information superhighway.

In Canada, because of higher long-distance rates and lower local rates, one would expect long-distance revenues to compose a greater proportion of total revenue than in the United States. However, the data in table 1-2 show the opposite: Canadian telephone

52. The real capital stock of the U.S. radio and television industry (including cable television) has grown substantially in the past twenty years. The United States has also allowed large telecommunications users to build their own private networks, but Canada has not allowed private entities to obtain microwave licenses for this purpose.

Table 1-3. U.S. Long-Distance Carrier Revenues, 1984 and 1993
Billion U.S.$

Company	*1984*	*1993*
AT&T	34.9	35.7
MCI	1.8	10.9
Sprint	1.1	6.1
Local telephone companies	12.4	13.8
Others	0.9	8.5
Total	51.1	75.0

Source: FCC, *Statistics of Communications Common Carriers*, 1993/94, p. 7.

companies derive nearly 43 percent of their revenues from local services, while U.S. telephone companies derive only 37 percent from these local services. This anomaly is the result of key differences between the United States and Canada. First, the number of long-distance calls in Canada is substantially below that in the United States. Second, the percentage of the population with telephones is higher in Canada (thus generating more local revenue). Third, extended-area service is much more prevalent in Canada than in the United States, resulting in a larger share of "local" calls in total calling.

The data in table 1-2 exclude the non-telephone company cellular carriers, the independent fiber-optics networks in large U.S. metropolitan areas, and private carriers. There are no comprehensive data on these unregulated carriers, but they still do not have a large share of the telecommunications market. As cellular grows and cable television systems enter the switched-services markets, the relative importance of traditional carriers will undoubtedly decline.

Long-Distance Carriers

In the United States, AT&T, the largest carrier, has seen its share of long-distance revenues decline steadily since it divested its local operating companies as a result of the antitrust suit in 1984 (see table 1-3). AT&T's two largest competitors are MCI, with nearly 15 percent of long-distance revenues, and Sprint, with 8 percent.[53]

53. See FCC, *Statistics of Communications Common Carriers*, 1994–95 edition, p. 7. These shares are for all long-distance services, including the short-distance intra-LATA services that are still largely the domain of the local telephone compa-

AT&T now has less than 50 percent of *all* long-distance revenues, intrastate and interstate, but about 60 percent of interstate revenues. According to FCC data on switched access minutes, AT&T's interstate share has fallen from 84.2 percent in the third quarter of 1984 to less than 60 percent in 1994.[54]

Although the U.S. interexchange market is dominated by three large carriers, the smaller long-distance carriers have been growing quite rapidly. Firms such as LDDS/Worldcom and Cable and Wireless are among the most important of these smaller carriers, accounting for more than $4 billion in revenues in 1994.[55] In addition, there are literally hundreds of resellers who purchase wholesale services from the facilities-based carriers and resell these services to smaller customers. According to the FCC, there were more than 800 carrier codes in existence in 1994 and almost 400 different carriers who ordered equal access from local telephone companies in order to connect their customers.[56] Most of these carriers are resellers, who may own a limited amount of switching equipment, but they generally do not have large transmission networks.

In Canada, a national long-distance carrier never developed. Instead, the planning of interprovincial facilities, connections, and pricing has been carried out by an association of provincial telephone companies, formerly called Telecom Canada.[57] With Unitel's entry, first as a reseller and then in 1992 as a facilities-based long-distance carrier, the members of Telecom Canada formed a new association, Stentor, to provide a unified national long-distance service. Because of its very recent entry, Unitel has a small share of the long-distance market, although it does have 25 percent of the private-line market. Various estimates exist for the market shares

---

nies. MCI and Sprint have 17 percent and 10 percent, respectively, of all long-distance carriers' (excluding the local companies) revenues.

54. Because AT&T and other long-distance carriers originate and terminate some traffic over leased private lines, it is difficult to estimate total interstate use with precision. The data on switched minutes reflect only that traffic that is switched by local exchange carriers (FCC, *Statistics of Communications Common Carriers,* 1993–94 edition, p. 311).

55. FCC, *Statistics of Communications Common Carriers,* 1994–95 edition, p. 7.

56. FCC, *Statistics of Communications Common Carriers,* 1993–94 edition, pp. 313–15.

57. Bell Canada operates in both Ontario and Quebec. Most other provinces have a single dominant telephone company and a number of smaller companies.

for the players in the Canadian long-distance market. The numbers are unreliable except in the Bell Canada and B.C. Telephone territories, where the CRTC has had jurisdiction for a long time. Stentor had 89 percent of the overall Canadian long-distance market in 1992, 86.3 percent in 1993, and less than 80 percent at present. Unitel's share has risen from 5.2 percent in 1992 to 7.2 percent in 1993 and more than 10 percent today. Resellers accounted for 4.9 percent in 1992 and 6.5 percent in 1993.[58] These data likely contain private-line revenues, as do the U.S. data.[59]

Local Telephone Companies

In the United States the seven regional Bell operating companies continue to account for 75 percent of local-company revenues and 79 percent of local revenues (table 1-4). Most of the rest is controlled by operating companies owned by two other holding companies—Sprint and GTE, which now own the Contel, Central, and United telephone companies—and a few other large independents, such as Southern New England Telephone, Pioneer (formerly Rochester Telephone), Alascom, and Lincoln (Neb.) Telephone. In 1984 the FCC decided to reprice telephone service by imposing a monthly flat-rate subscriber line charge that was to replace part of the interstate carrier access charge that was levied on long-distance companies to connect their calls. The result may be seen in the decline since 1984 in the share of revenues accounted for by carrier access revenues, particularly those of the Bell companies. Perhaps more surprising is the slow growth in local-company revenues from all sources combined. Between 1984 and 1993, total U.S. local carrier revenues failed to grow as rapidly as the GDP deflator, and revenues *per line* were essentially constant in nominal dollars. Note, however, that independents' revenues grew much more rapidly than those of the Bell companies, largely because their access lines grew much more rapidly.

58. Response of Stentor Research Center, Inc., to Interrogatory, SRCI (CRTC) 31 May 93-102 RRF Supplemental, pp. 13–20. Other competitors of Stentor are estimated as having an additional 8 percent of the private-line market, leaving the Canadian telephone companies with 67 percent.

59. FCC, *Statistics of Communications Common Carriers*, 1993–94 edition, p. 7.

Table 1-4. Revenues of U.S. Local Telephone Companies, 1984 and 1993
Billion current U.S.$ (except revenues per line)

Type of company	*1984*	*1993*	*Percentage change*
Bell operating companies			
Local services	26.6	39.8	49.6
Carrier access services	16.2	15.7	−3.1
Long-distance services	9.0	9.8	8.9
Miscellaneous	6.2	6.6	6.5
Total BOCs' revenues	58.0	72.0	24.1
Total BOC revenues per access line	634	627	−1.1
Independent companies			
Local services	5.9	10.8	83.1
Carrier access services	4.7	6.8	44.7
Long-distance services	3.3	3.7	12.1
Miscellaneous	1.6	3.2	100.0
Total independents' revenues	15.5	24.5	58.1
Total independent revenues per access line	674	725	7.6
Total U.S. local carrier revenues	73.5	96.5	31.3
Total U.S. carrier revenues per access line	642	649	1.1
Implicit GDP deflator	91.0	123.5	35.7

Sources: FCC, *Statistics of Communications Common Carriers*, 1984 and 1993/94 (Bell companies); USTA, *Statistics of the Local Exchange Carriers*, 1984 and 1993 (independents); U.S. Department of Commerce (GDP deflator).

The large U.S. telephone companies are heavily involved in foreign telephone operations, generally as partners with non-U.S. firms. British Telecom has purchased 20 percent of MCI, and MCI has invested in Canada's second largest long-distance carrier, Unitel.[60] France Telecom and Deutsche Telekom are proposing to purchase 20 percent of Sprint and to develop an international consortium among all three companies. NYNEX and SBC (formerly Southwestern Bell) have equity interests in British cable television companies. Bell Atlantic is a part owner of New Zealand Telecom. Bell South has wireless operations in New Zealand and Australia. And several RBOCs have interests in eastern European telephone companies. Despite all of these ventures, however, most of the large U.S. local-exchange companies continue to derive a large share of their revenues from U.S. local voice/data communications, a market that is declining in real terms.

60. *Sprint Corporation* v. *American Telephone & Telegraph Corporation*, File No. ISP-95-002, November 18, 1994, p. iii.

Table 1-5. Estimated Total Operating Revenues for Telecommunications Carriage Industry in Canada, 1993
Billion Cdn.$

Carrier	*Total operating revenues*
Stentor companies	
AGT (Alberta)	1.16
BC Tel	2.20
Bell Canada (Ontario & Quebec)	7.91
Island Tel (PEI)	0.05
Manitoba Tel	0.53
Maritime Tel & Tel (Nova Scotia)	0.47
New Brunswick Tel	0.35
Newfoundland Tel	0.27
Saskatchewan Tel	0.55
Telesat Canada	0.20
Quebec Telephone	0.24
Total Stentor	13.89
Independent telephone companies (50)	0.90
Long-distance companies	
Unitel, Sprint Canada, Westel, Teleroute	0.60
Overseas carrier	
Teleglobe Canada	0.40
Resellers	0.20
Radio common carriers	
Including cellular, paging, and other wireless carriers	1.40

Source: Industry Canada, *Telecommunications Service Industries, 1994–95* (November 1994), p. 16.

Because entry liberalization came much later to Canada than to the United States, the large carriers continue to dominate the Canadian industry. As table 1-5 shows, the Stentor group has 87 percent of all Canadian telecom revenues excluding cellular and radio common carriers.

Cellular Service

Cellular service began in the United States in 1983, but by 1994 U.S. cellular companies had more than 23 million subscribers, or one subscriber for every eleven persons in the country (table 1-6). This phenomenal growth is projected to continue for the rest of this decade, and cellular subscribers are forecast to grow to more

Table 1-6. Subscribers to Cellular Service in Canada and the United States, 1990–94

Millions; percent of population in parentheses

	1990	*1991*	*1992*	*1993*	*1994*
Canada	0.6 (2.2)	0.79 (2.8)	1.13 (4.0)	1.43 (4.9)	1.99 (6.8)
United States	5.3 (2.1)	7.6 (3.0)	11.0 (4.3)	16.0 (6.2)	23.6 (9.1)

Source: Canada: Organization for Economic Cooperation and Development, *Communications Outlook* (1995); and personal communication with Cantel.

United States: Cellular Telephone Industry Association, annual surveys.

than 80 million by the year 2000.[61] Because the service began somewhat later in Canada, the Canadian cellular industry has a lower penetration.

In both countries cellular service is offered by two facilities-based rivals. The United States carriers are licensed on a market-by-market basis; the two Canadian services are national in scope. However, the ten largest U.S. cellular companies, including AT&T, GTE, Sprint, and the seven RBOCs, account for about three-fourths of all U.S. subscribers. Recently a third carrier, Nextel, has entered the U.S. cellular market, offering a combination cellular-paging service, but technical problems have slowed its progress. In addition, the United States has just finished its first major auction of two new PCS spectrum licenses for each major trading area.[62] Thus, the United States will soon have as many as five competing wireless carriers in each market.

Telephone Penetration

Underlying most of today's telecommunications regulation is the desire to promote universal service, a major raison d'être for telephone regulation shortly after the turn of the century.[63] For a variety of reasons, regulators have traditionally used long-distance

61. Liebowitz and others, *The Wireless Communications Industry.*

62. In three areas—New York, Los Angeles–San Diego, and Washington-Baltimore—only one license was auctioned because a second had already been granted to firms before the auction under the FCC's pioneer preference policy.

63. There is considerable controversy about this episode in economic history. For opposing views, see Milton Mueller, *Universal Service: Competition, Interconnection, and Monopoly in the Making of the American Telephone Industry* (American Enterprise Institute, 1995); and Temin, *The Fall of the Bell System.*

Table 1-7. Percentage of Households with Telephone Service in the United States and Canada, Selected Years, 1960–92

Year	*Canada*	*United States*
1960	83.3	77.0
1970	93.9	87.0
1980	97.6	93.0
1985	98.2	91.8
1990	98.5	93.3
1992	99.0	93.9

Sources: Canada: *Statistics Canada;* "Household Facilities by Income and Other Characteristics," various years. United States: For 1960–80, U.S. Bureau of the Census, *Census of Housing*; for 1985–92, U.S. Bureau of the Census, *Current Population Survey*.

and other revenues to cross-subsidize local access.[64] For this reason and because of the very high level of per capita incomes in the United States and Canada, a large share of households subscribe to telephone service (see table 1-7). However, the United States has lagged behind Canada in residential telephone penetration despite the slightly higher income levels in the United States.

A closer look at telephone service across demographic groups provides some insight into the differences between the United States and Canada. Table 1-8 shows that poor U.S. households are much less likely to have telephone service than are poor Canadian households. Fewer than 85 percent of poor U.S. households have telephone service, but about 97 percent of poor Canadian households have a telephone. The differences are much narrower for households with above-average income.

The differences in telephone subscribership between the two countries may be due in part to the larger share of poor young households in the United States, particularly those headed by a young single parent. This is undoubtedly a reflection of the more generous income-maintenance system and the existence of state-financed health care in Canada. Income (after taxes and government-provided services) and transfer payments are higher for poor Canadians than for poor U.S. citizens. Table 1-8 also reveals that the Canadian-U.S. difference in telephone subscription for households headed by someone twenty-six years of age or younger is

64. The consumption externalities in communications provide the efficiency rationale for subsidizing connections to the telephone network, but this rationale is of much less importance in countries, such as Canada and the United States, in which virtually everyone has or has access to a telephone.

Table 1-8. Percentage of Households with Telephone Service by Demographic Group, Canada and the United States, 1992

Group	*Canada*	*United States*
All households	99.0	93.8
Income group		
<$10,000	93.6	80.6
$10,000–20,000	98.0	92.0
$20,000–40,000	99.2	97.3
>$40,000	99.8	99.4
Low-income cutoff[a]		
At or above	99.6	97.6
Below	96.7	84.4
Age of household head		
<25 years (< 35)[b]	98.3	82.0
25–64 years (35–65)[b]	99.0	93.6
>64 years (> 65)[b]	99.4	97.3

Sources: Statistics Canada 13-218, "Household Facilities by Income and Other Characteristics, 1992," author's calculations; U.S. Bureau of the Census, Poverty Division, *Current Population Survey*; Alexander Belinfante, "Telephone Subscribership in the United States," *FCC News*, December 1993.

a. For both one-person and single-family households.

b. Canada.

around 16 percentage points, while the differences are much narrower for older households.

Another explanation for the higher penetration rates in Canada may be the lower residential access rates in Canada. Canadian residential rates are between 16 and 34 percent below U.S. rates,[65] but long-distance rates are much higher in Canada. It is possible that this difference in rates translates into an almost 2 percentage-point difference in residential subscriber penetration,[66] or perhaps 40 percent of the overall average 5 percentage-point differential. Finally, as Lewis Perl has pointed out, cold-weather states have higher telephone penetration than warmer states. Our own results suggest that this may account for as much as 1.5 percentage points of the difference.[67]

65. This is an average. See table 6-4.

66. This is at best a crude estimate based on the assumption that the price elasticity of demand for access is about minus 0.05. A recent study concludes, however, that the demand for access is also an inverse function of long-distance rates. (See Jerry Hausman, Timothy Tardiff, and Alexander Belinfante, "The Effects of the Breakup of AT&T on Telephone Penetration in the United States," *American Economic Review*, vol. 83 (May 1993), pp. 178–84. Thus, it is far from clear that the Canada-U.S. differences in subscribership are due to rate differences.

67. Specifically, a loglinear regression analysis of penetration on access prices, per capita income, share of the population outside metropolitan areas, and geographical location finds that southern location reduces telephone subscriptions by 3 per-

There are other differences between the U.S. and Canadian telecommunications industries. In Canada, noncommunications companies are not able to obtain microwave licenses. In the United States private microwave has been allowed since 1959. One Canadian carrier, Teleglobe, has a monopoly on international traffic, other than U.S.-Canada traffic, while several U.S. carriers are now actively battling for international traffic.

As mentioned earlier, the differences between the Canadian and U.S. telephone industries may now be narrowing for a number of reasons. First, cross-border alliances and cross-ownership are growing as Canada and the United States continue to liberalize and in response to the Canada–U.S. Free Trade Agreement. Second, the Canadian regulatory regime is following the U.S. precedent in moving toward greater liberalization (see chapter 2). Third, the same two companies—Northern Telecom and AT&T—dominate the large telephone switch-gear market in both countries; thus the technology in the two countries is similar. Fourth, AT&T and MCI have recently entered into strategic alliances with Unitel and Bell Canada, respectively.[68] How the two countries evolve in the next few years depends on how regulators allow competition, including the growing competition between telephone and cable.

Summary

The telephone industries of the United States and Canada, and indeed of much of the rest of the world, are undergoing a fundamental change driven by an explosion of new technologies and a questioning of the old order. State-controlled monopolies are being replaced by regimes featuring competitive rivalry. New services are being developed by telephone companies, cable television companies, cellular providers, and a variety of other new suppliers. Regulators in Canada and the United States are not unaware of these developments, and they are now attempting to fashion poli-

---

cent. If half of the U.S. population lives in northern areas much like Canada, the difference due to weather or geography may be about 1.5 percentage points.

68. Sprint has also formed an alliance with an erstwhile Canadian reseller that is now called Sprint Canada.

cies that will allow regulated carriers to offer a mixture of competitive and regulated monopoly services.

The remainder of this book will examine the performance of the regulators and regulated telephone companies and the effects of regulation on both sides of the border. The focus will be on conventional telephone service because such service is still more than 90 percent of the telephone companies' business. This examination will show that the regulation of this traditional business is likely to be responsible for slowing the transition from POTS to a more dynamic industry, delaying the offering of the myriad of new services that are likely to emerge from the revolution in electronics and communications technologies. Increasingly, talk is cheap, but regulators defend the status quo to perpetuate a system of cross subsidies and to protect their chosen instruments for delivering these subsidies, the incumbent telephone carriers.

Chapter Two

Regulating the Telephone Industries

THE RAPID CHANGES IN telecommunications technology have placed regulators in many countries in a difficult position. They are increasingly unable to resist the pressure to admit new players and to allow new service offerings from new and old firms alike, but they are unable to stand aside and let the market sort out the winners and losers. Accustomed to the quiet life of simply ruling on the adequacy of returns for telephone carriers, these regulators are drawn into passionate disputes about the fairness of competition and the ability of regulated carriers to satisfy various social responsibilities, including the maintenance of below-cost access to the network for numerous subscribers.

Nowhere are these tendencies more visible than in the United States and Canada, where regulated competition (an oxymoron at best) is steadily replacing the staid old regulated monopolies.[1] In both countries there are continuing disputes over the degree to which regulated monopolists should be allowed to compete in other regulated and unregulated businesses and on what terms. Telephone companies have been barred from offering traditional cable television services, and cable television companies in most jurisdictions have not yet received approval to offer telephony services. In both countries telephone and cable rates are subject to federal regulation, and in the United States telephone services are subject to state regulation. These regulatory policies are increasingly viewed as outmoded and as providing disincentives for allocative efficiency. As a result, the Canadian and U.S. regulators are under strong pressure to reform their regulatory programs.

1. Similar pressures exist in the United Kingdom and Australia. In New Zealand, total deregulation has thrust the regulatory issues into the courts.

Regulatory Authority in the United States

There is a widespread misconception that U.S. telecommunications has been deregulated and that Canadian authorities are following the U.S. lead toward unfettered competition in telecommunications. In the United States, liberalized entry and the AT&T divestiture are correctly associated with a freer market, but regulation remains for most services. Interstate services are still regulated by the FCC. Intrastate services continue to be regulated by state public utility commissions. Only terminal equipment has been completely deregulated and, for the most part, detariffed.[2] Inside wiring is being detariffed. And many states now allow companies considerable rate flexibility for nonessential, competitive service offerings.

The United States has a long history of economic regulation of transportation, communications, and finance at both the federal and state levels. The first major federal regulatory agency was the Interstate Commerce Commission (ICC), established in 1887 to regulate interstate surface transportation—at first, the railroads. In 1910 the Mann-Elkins Act, enacted in response to complaints that AT&T was engaging in monopolistic practices in denying independent telephone companies access to its long-distance network, extended the ICC's jurisdiction to interstate telephone service.[3]

Jurisdictional Tensions

In 1934, Congress passed the Federal Communications Act, establishing a new Federal Communications Commission (FCC) entrusted with the responsibility of rationing the electromagnetic spectrum and regulating foreign and interstate communications services. The FCC was expressly forbidden to regulate intrastate communications services, but subsequent court decisions have limited this exclusion to those matters that are " . . . separable from

2. Detariffing is simply the removal of the requirement that carriers must file tariffs with the regulatory authority for the given service.

3. For different views of this history see Peter Temin (with Louis Galambos), *The Fall of the Bell System* (Cambridge University Press, 1987); and Milton Mueller, *Universal Service: Competition, Interconnection, and Monopoly in the Making of the American Telephone Industry* (American Enterprise Institute, 1995).

and do not substantially affect the conduct or development of interstate communications."[4] As a result the FCC has been able to assert regulatory authority over the measurement and division of the rate base, inside wiring, and terminal equipment.[5]

The dual system of regulating telephone service at the state and federal levels has always been accompanied by a certain degree of tension between the state regulators and the FCC. The 1934 act requires the establishment of a federal-state joint board to make recommendations regarding the appropriate separation of property and expenses between the state and federal jurisdictions for rate-making purposes, but the FCC has final authority in such matters. Whatever the jurisdiction, until the early 1980s, regulators used a rate-of-return, rate-based mechanism. This mechanism was designed to ensure that investors' private property was not expropriated through a regulatory mechanism. Rates were to be set so as to provide a satisfactory rate of return on the capital invested in the telephone company. This regulatory mechanism is now considered by most industry participants, academics, and a growing number of regulators to be flawed, expensive, and inefficient.[6] New regulatory mechanisms are now replacing rate-of-return regulation in the United States at the FCC and in a substantial number of states.

In the past twenty-five years the FCC has generally promoted a slow transition to competition and the movement of rates toward costs, while the state regulatory commissions have been reluctant to move away from protecting the carriers' monopoly status and the rate distortions made possible by that monopoly, namely, low monthly rates for flat-rate local residential service paid for by prices that are above costs for many other services. For the most part the FCC has been able to preempt the states in moving toward liberalization and market competition. The commission won a sweeping victory in 1976 when the U.S. Court of Appeals affirmed the FCC's right to mandate competition in equipment on the customer's premises even though such equipment was used more than

4. *North Carolina Utility Commission* v. *FCC,* 537 F.2d 793 (4th Cir. 1976).

5. The extent of the FCC's preemption authority is currently subject to some debate given the Supreme Court's 1986 decision in *Louisiana Public Service Commission v. FCC,* 476 U.S. 355 (1986), in which the Court invalidated an FCC ruling on depreciation rules for telecommunications equipment in the intrastate jurisdiction.

6. These shortcomings are discussed more fully in chapter 4.

90 percent of the time for intrastate calls.[7] Subsequently, the FCC's decision to deregulate this equipment on the customer's premises altogether and to preempt all state attempts to regulate it was upheld by the courts.[8] However, the trend toward preemption of state regulation by the FCC was abruptly halted by the Supreme Court's 1986 rejection of an FCC attempt to prescribe depreciation rules for telecommunications equipment.[9]

The limits to federal preemption of state telecommunications regulation are essentially provided by the 1934 Communications Act. Michael Kellogg, John Thorne, and Peter Huber note that an FCC preemption order is likely to be upheld if it serves the valid goals brought under FCC jurisdiction by the Communications Act and is necessary for achieving these goals.[10] But the FCC probably cannot, under current law, *require* states to admit entrants into local access/exchange service. For this reason, 1994–95 proposals to reform telecommunications policy often include provisions for preempting state regulators' attempts to maintain regulated local monopolies.[11]

State Regulation

In describing recent state regulatory developments, it is useful to distinguish interexchange (long-distance) services from other services. The United States, as a result of the federal system of government and the AT&T divestiture, now has three distinct levels of interexchange service. Intra-LATA service is long-distance service provided within the local access and transport areas (LATAs) into which the country was divided by the AT&T decree. Some states have more than one LATA; therefore, calls between LATAs within a single state are intrastate, inter-LATA calls. Finally, calls between states are interstate, (generally) inter-LATA calls. As a result of the AT&T decree, substantial pressure has been brought

7. *North Carolina Utility Commission* v. *FCC,* 537 F.2d 787 (4th Cir. 1976).

8. *Computer and Communications Industry Association* v. *FCC,* 693 F. 2d 198 (D.C. Cir. 1982).

9. *Louisiana Public Service Commission* v. *FCC,* 476 U.S. 355 (1986).

10. Michael K. Kellogg, John Thorne, and Peter W. Huber, *Federal Telecommunications Law* (Little, Brown and Co., 1992), pp. 111–12.

11. H.R. 1555 and S. 652 introduced in the 104th Congress provide for such preemption.

on state commissions to allow all interexchange carriers—AT&T, MCI, and Sprint, for instance—to compete in intrastate inter-LATA and even in intra-LATA markets. Liberalization of these markets raises issues of (symmetric or asymmetric) long-distance rate regulation.

LOCAL TELEPHONE SERVICE. In the early 1980s the state commissions were faced with a myriad of rate filings for local rate increases.[12] These rate-increase filings followed a period of rate suppression during the inflationary 1970s and were exacerbated by the FCC's decision to rebalance rates by instituting a monthly subscriber line charge for all telephone customers and reducing interstate long-distance rates.[13] These local rate increases led to severe criticism of state regulatory officials. As a result, they began to search for mechanisms that might induce improvements in carrier efficiency and, therefore, lower rates in the future.

Most states have traditionally used some form of rate-of-return regulation in response to court edicts that rates must be just and reasonable.[14] In practice this type of regulation amounts to a periodic review of rates or rate proposals based on an evaluation of the adequacy of a carrier's return on capital. Because rate-of-return regulation reduces the incentive for carrier efficiency, provides an incentive for excessive capitalization, and may create the opportunity for cross-subsidizing competitive forays from monopoly rev-

12. The FCC reports that state telephone rate cases involved more than $4 billion in requested revenue increases in 1984. In recent years, these requests have fallen to less than $0.5 billion a year. See periodic Federal-State Joint Board, *Monitoring Reports* (FCC, CC Docket No. 87-339), p. 389, table 5.14.

13. The Federal Communications Commission and the states apportion between the federal and state jurisdictions the costs of local telephone company operations. Traditionally the federal component of the fixed, non-traffic-sensitive costs was recovered entirely through interstate long-distance rates, not through flat monthly charges on end users (subscribers). This practice resulted in inefficiently high interstate long-distance rates, a result that the FCC began to address in 1984 by assessing part of the non-traffic-sensitive costs to end users on a flat monthly basis and reducing the access charges levied on long-distance companies. To the typical subscriber, this shift had the appearance of increasing the local monthly bill for flat-rate access. Those subscribers who make a large number of interstate calls may have noticed a substantial decline in their long-distance charges, but those who do not typically place many interstate calls would not have received an offset to the increase in monthly charges, which have now reached $42 a year for residences and as much as $72 a year for multiline businesses.

14. See Kellogg and others, *Federal Telecommunications Law,* ch. 9, for a discussion of this case law.

Table 2-1. United States: State Regulatory Reform for Local Exchange Carriers, 1994

Deregulation	*Tiered deregulation*	*Revenue sharing*	*Price caps*	*Rate freeze*	*None*
Idaho[a]	Arizona	Alabama	California	Kansas	Arkansas
Nebraska	California	California	Michigan	Maryland	Connecticut
	Colorado	Colorado	New York (limited)[b]	New Jersey	Delaware
	Idaho	Florida	North Dakota	Oregon	Indiana[c]
	Illinois	Georgia	Oregon (limited)	Texas	Massachusetts
	Iowa	Idaho	Rhode Island	Vermont	Missouri
	Maine	Kentucky	Washington (limited)	West Virginia	New Hampshire
	Maryland	Louisiana	Wisconsin (limited)		New Mexico
	Michigan	Maryland			North Carolina
	Minnesota	Minnesota			Ohio[c]
	Montana	Mississippi			South Carolina
	Nevada	Nevada			Utah
	New Jersey	New Jersey			
	North Dakota	New York[b]			
	Oklahoma	Oregon			
	Oregon	Rhode Island			
	Pennsylvania	Tennessee			
	South Dakota	Texas			
	Vermont	Washington			
	Virginia				
	West Virginia				

Source: Telecom Publishing Group, *State Telephone Regulation Report* (Alexandria, Va., February 10 and 24, 1994).
a. All but basic local service.
b. Separate regime for Rochester Telephone.
c. For large companies only.

enues, state commissions have further reasons to look for alternative mechanisms for regulating their local-exchange carriers.

Unfortunately, this search has yielded few new ideas. The most common new regulatory options now in use or under study in various states are tiered regulation, revenue sharing, and price caps. Tiered regulation is simply the establishment of different degrees of regulation for various service groups (or tiers) depending on the perceived vigor of actual or potential competition in the respective service. The most competitive service categories, such as terminal equipment, may be totally deregulated. The next tier may be services such as inter-LATA long-distance or various central-office "customized" offerings in which entry has been permitted but which are not yet subject to vigorous competition. This tier may be subjected to limited regulation, particularly for the carrier perceived as dominant. The highest tier, requiring the most regulatory supervision, would include the monopoly services such as basic local access.

Tiered regulation, then, is little more than selective deregulation or regulatory relaxation. Since formal rate-of-return regulation generally remains for monopoly services, few of the problems associated with the rate-of-return mechanism are avoided. Despite this shortcoming, twenty-one states had adopted a tiered approach as of 1994, and many others were examining it (see table 2-1).[15]

Revenue sharing is a modest improvement on rate-of-return regulation, allowing carriers to keep a share of profits in excess of their allowed return on capital. Usually the sharing criterion is based on an accounting rate of return. If a carrier earns more than its allowed or targeted rate, it is permitted to keep a share of the excess up to a maximum level—presumably less than unconstrained monopoly equilibrium profits. Nineteen states had recently adopted this form of regulation as of 1994 (see table 2-1).

Price caps are also beginning to find acceptance as a partial substitute for rate-of-return regulation. California and seven other states had adopted some form of price caps as of 1994, and several others were considering a similar approach. Under price cap regulation, firms are allowed to raise average rates by a maximum

15. A few others have adopted various rate flexibility provisions for competitive services. Table 2-1 shows only those that have deregulated the most competitive services.

rate per year, generally defined as the inflation rate less the industry's assumed improvement factor—a factor that depends on the firm's historical productivity growth. Individual service rates may be rebalanced as long as the average rate does not rise by more than the price cap. To cushion the effects on politically sensitive service rates, the maximum annual increase or decrease in individual rates may also be limited even if the *average* rate remains within the allowed rate cap.

Only one state, Nebraska, has deregulated all telecommunications services, but this deregulation is severely constrained by four provisions of the legislation that launched it in 1986. First, entry is prohibited in local access-exchange services. Second, rate increases for local service are subject to Public Service Commission review if they rise by more than 10 percent a year. Third, all rate changes must be announced 120 days in advance. Finally, all long-distance rates must continue to be set solely on the basis of mileage, not density.[16]

Few other states have allowed any measurable degree of deregulation of basic intrastate telecommunications services. Idaho has deregulated all services except basic access-exchange service for customers with fewer than five lines. Several others have deregulated inter-LATA long distance, but few have deregulated intra-LATA toll. Only six states allowed competitive entry into basic, switched local services as of 1994, but by mid-1995, this number had swelled to 16.[17]

A promising new version of regulatory reform has recently emerged in New York and in the Ameritech (midwestern) region. In 1992, Rochester Telephone petitioned the New York Public Service Commission for authority to restructure its operations into a holding company that would own, among other operations, a regulated company offering wholesale services and a less-regulated set of companies operating in competitive markets. One of the latter would resell local service, purchased from the wholesale company, to retail customers. The wholesale company, R-Net, would

16. Milton L. Mueller, *Telephone Companies in Paradise: A Case Study in Telecommunications Deregulation* (New Brunswick, N.J.: Transaction Publishers, 1993).

17. Telecom Publishing Group, *State Telephone Regulation Report*, vol. 12, no. 12, and vol. 12, no. 13 (June 16 and 30, 1994), pp. 3–6 and 3–4, and vol. 13, no. 16 (August 24, 1995), p. 1.

be operated by a completely separate management subject to an independent board of directors. Other resellers could also purchase these wholesale services from Rochester. The wholesale company would unbundle its local-access services, offering switching, transport, subscriber-line service, and various other services under tariff to any and all competitive resellers. Thus an entrant could purchase its own switch, build its own local trunks, and connect to some or all of Rochester's subscriber lines.

In 1994 the New York Public Service Commission finally approved the Rochester plan, but only with substantial modifications.[18] Rochester would be forced to continue to offer retail services to subscribers through its regulated wholesale subsidiary, but it could also offer bundled services through its competitive resale company. Despite early start-up problems, it appears that new local carriers are entering the Rochester market as resellers and even facilities-based carriers.

Ameritech offered a similar plan to its state regulators and to the U.S. Department of Justice. Ameritech would encourage its state regulators to allow entry into local access-exchange services, unbundle its network, ensure number portability, and provide other competitive safeguards in return for being allowed to enter the inter-LATA market in its five-state area. As of April 1995, Ameritech had been forced to limit its plan to an experiment that would extend only to part of Michigan and the Chicago LATA. Moreover, Ameritech would only be allowed to enter the inter-LATA market as a reseller. Nevertheless, the Rochester and Ameritech plans represent the first major attempt to unbundle and liberalize local access.[19]

Many of the states that have enacted some form of regulatory reform or liberalization have also required limited—one- or two-year—rate freezes on some array of basic services. Presumably these freezes were the political price that the carrier had to pay to persuade state electorates of the wisdom of a new set of less intrusive regulatory rules. In at least one state, New York, this freeze proved disastrous for the regulated carrier, New York Telephone,

18. See *State Telephone Regulation Report*, vol 12, no. 21 (October 20, 1994), pp. 4–5.

19. One of the authors (Crandall) worked on behalf of both Rochester and Ameritech in these regulatory proceedings.

Table 2-2. United States: Intrastate Regulation of AT&T by the States, 1994

No rate of return regulation for AT&T			*Rate of return regulation for AT&T*
With full pricing flexibility	*With floor and/ or ceiling rates*	*With streamlined rate review*	
Idaho	Alabama	Connecticut	Arizona
Illinois	California	Oklahoma	Arkansas
Indiana	Colorado	Rhode Island	Wyoming
Iowa	Delaware	West Virginia	
Kentucky	Florida		
Louisiana	Georgia		
Maine	Kansas		
Maryland	Michigan		
Massachusetts	Mississippi		
Minnesota	Missouri		
Montana	Nevada		
Nebraska	New Mexico		
New Hampshire	New York		
New Jersey	North Carolina		
Oregon	North Dakota		
Pennsylvania	Ohio		
South Dakota	South Carolina		
Texas	Tennessee		
Utah	Wisconsin		
Vermont			
Virginia			
Washington			

Source: *State Telephone Regulation Report*, October 20 and November 3, 1994.

since it subsequently fell far short of the lower range of its revenue-sharing formula. As a result, New York reverted to traditional rate-of-return regulation of New York Telephone.

LONG-DISTANCE SERVICES. The states' approach to regulating intrastate long-distance service is also changing. Inter-LATA toll competition is allowed in virtually all of the thirty-eight multi-LATA states. In each of the states with inter-LATA competition, the significant policy issues involve the regulatory treatment of AT&T. In all but three states, AT&T is no longer subject to formal rate-of-return regulation (table 2-2).[20] However, this does not mean that AT&T has complete rate flexibility in all of these states. In nineteen of these states, AT&T is subject to some form of banded rate flexibility with prescribed floors, ceilings, or both to allow rate changes,

20. Table 2-2 contains information for some single LATA states, such as Rhode Island, in which AT&T operates.

and in another four states all rate changes are subject to streamlined review.

In intra-LATA markets, forty-four of the forty-eight mainland states allow facilities-based competition, but of these only six have fully deregulated intra-LATA rates (table 2-3), or at least streamlined regulation to allow prompt approval of rate increases or decreases. As of late 1994, however, no state required that the local-exchange carriers provide equal-access 1+ dialing service for all service offered by competitors.[21] As a result, a short-distance intrastate toll call (within a LATA) that is dialed simply as 1+ the seven-digit local number automatically goes over the local exchange company's network. For this reason, intra-LATA competition may not be very strong even though a state commission allows facilities-based interexchange (IX) carriers such as MCI, Sprint, and AT&T to offer such service.

Federal Regulation

Until the mid-1960s federal regulation of interstate telephone services was largely an informal supervision of AT&T rates. Little attention was given to the structure of rates, and AT&T was only required to file tariffs that would result in its meeting a maximum rate-of-return standard. Occasionally disputes would arise about individual rates, such as interconnection rates for broadcasters, but AT&T was otherwise relatively free to determine its own rate structure.

The advent of microwave and the resulting real declines in long-distance costs began to change this comfortable relationship. First the FCC decided in 1959 to allow private users to build their own microwave networks. At about the same time, the FCC and state regulators agreed to begin increasing the share of AT&T's costs allocated to interstate activities—even though the cost of long-distance services was falling relative to local access and exchange costs—thereby preventing long-distance charges from falling as rapidly as they should and keeping local rates artificially low. The result of these two decisions was to invite entry into long-distance

21. This situation is apparently about to change. Many state commissions are considering ordering that the local carriers provide 1+ equal access to all intra-LATA competitors.

Table 2-3. United States: State Regulation of Local Exchange Companies' Intra-LATA Long-Distance Service, 1994

States allowing facilities-based competition			*States allowing no competition*	
With full rate regulation	*With floor/ ceiling rates*	*With full rate flexibility*	*With full rate regulation*	*With floor/ ceiling rates*
Alabama	Connecticut	Idaho[a]	New York	Arizona
Arkansas	Delaware	Maryland	Virginia	California[c]
Colorado	Georgia	Minnesota[b]		
Florida	Illinois	Nebraska		
Iowa	Indiana	South Dakota[b]		
Kentucky	Kansas	West Virginia[b]		
Louisiana	Michigan			
Maine	Mississippi			
Massachusetts	Missouri			
Montana	Nevada			
New Mexico	New Hampshire			
North Carolina	New Jersey			
Oklahoma	North Dakota			
South Carolina	Ohio			
Texas	Oregon			
Utah	Pennsylvania			
Washington	Rhode Island			
Wyoming	Tennessee			
	Vermont			
	Wisconsin			

Source: *State Telephone Regulation Report*, October 20 and November 3, 1994.
a. Southern Idaho only.
b. Streamlined regulation.
c. Competition to begin January 1, 1995.

services by new microwave-based carriers. Indeed, the first such entrant appeared at the FCC's doorstep in 1963. After six years of petitioning the FCC, this entrant, MCI, was granted entry to provide limited private-line services, but the FCC and the states continued to shift the costs of AT&T's activities for rate-making purposes toward the interstate long-distance activities.

Throughout the 1960s and 1970s, the FCC became more and more convinced of the benefits of liberalized entry. Although the commission did not intend to admit competition into switched long-distance services as early as the mid-1970s, MCI entered the MTS market in 1974 by launching its switched Execunet service without FCC permission. After failing to stop MCI in the courts, the FCC essentially conceded to a brave new competitive world. By the end of the 1970s, the commission had concluded that competition in all interstate service markets and terminal equipment was in the public interest.[22]

Regulatory reform and liberalization have traditionally been much higher priorities of the federal government than of the state and local governments, which have often resisted federal attempts to reduce the degree of telecommunications regulation. The FCC led in the fight to reprice local access and toll rates, but the states, predictably, have been slow to follow. The states generally resisted the FCC's liberalization of terminal equipment. The FCC prevailed and subsequently even forced the unbundling and detariffing of embedded terminal equipment and inside wiring.[23]

NETWORK ACCESS. Faced with the convergence of computers and communications, the FCC enacted innovative rules in its *Computer II* decision[24] that were designed to let AT&T pursue competitive ventures while still offering regulated monopoly basic services. In that proceeding, the commission required that AT&T maintain structurally separate subsidiaries for its enhanced com-

22. The details of this history are found in Temin (with Galambos, 1987); Gerald Brock, *The Telecommunications Industry* (Harvard University Press, 1981); and Robert W. Crandall, *After the Breakup: The U.S. Telecommunications Industry in a More Competitive Era* (Brookings, 1991).

23. The FCC's attempt to detariff inside wiring was reversed in *NARUC* v. *FCC*, 880 F. 2d 422 (D.C. Cir. 1989), but inside wiring is still unbundled from the rest of local telephone service.

24. Amendment of Section 64.702 of the Commission's Rules and Regulations, Second Computer Inquiry, 77 FCC 2d 384; modified on recon., 84 FCC 2d 50 (1980).

petitive services, a requirement that was abandoned with the breakup of AT&T.

After the AT&T divestiture, the FCC concluded that the benefits of structural separation at the Bell operating companies (BOCs) and AT&T were outweighed by their costs. There was no easy way to identify the facilities offering basic services from those offering enhanced services. In addition, the concern over cross subsidies was somewhat counterbalanced by the concern that structural separation would deprive consumers of the benefits of economies of scope in telephone company operations. As a result, the FCC decided to eliminate its structural separation requirements for the BOCs, substituting a requirement for an unbundled open network architecture (ONA) through which the BOCs would be required to charge outside providers and their own affiliates the same rates for each network element offered.[25] Each of the BOCs has filed its own ONA plan, and predictably these plans have created controversy. ONA has expanded to include a requirement that local companies actually allow outside service providers to locate some of their interconnection facilities in the switching centers of the local-exchange carriers (LECs), but the requirement was reversed by the appellate court.[26]

REPLACING RATE-OF-RETURN REGULATION. In 1984 the United Kingdom enacted a price-cap form of regulation for the newly privatized British Telecom.[27] This form of regulation has attracted a wide following among academic students of the telecommunications industry, and it was embraced by FCC Commissioner Dennis R. Patrick as his principal goal for federal telecommunications regulation in 1985. After four years of political struggle, the FCC imposed a tightly constrained set of price caps on AT&T.[28] In 1990 price caps were also extended to the interstate operations (access

25. This was the result of the FCC's Third Computer Inquiry, announced in two separate decisions: Amendment of Section 64.702 of the Commission's Rules and Regulations, Report and Order, 104 FCC 2d 958 (1986) and 2 FCC Rec. 3072 (1987).

26. FCC, In the Matter of Expanded Interconnection with the Local Telephone Company Facilities, *Order*, 8 FCC Rec. No. 1 (1992). Reversed by *Bell Atlantic Tel. Cos.* v. *FCC*, 24 F.3d 1441 (D.C. Cir. 1994).

27. The genesis of price-cap regulation may be traced to Stephen C. Littlechild, "Regulation of British Telecommunications' Profitability," Report to the Secretary of State, Department of Industry, London, England, February 1983.

28. Policy and Rules Concerning Rates for Dominant Carriers, Report and Order and Second Further Notice of Proposed Rulemaking, 4 FCC Rec. 2873 (1989).

charges) of the Bell operating companies and the local exchange operations of GTE, but price caps were made optional for other local exchange carriers.

The AT&T price cap regulations established three baskets: one for residential and small business services, one for 800 services, and a third for all other business services (excluding customer-specific tariffs such as Tariff 12 or Tariff 15). Rates in each basket were to be allowed to increase at the rate of inflation less 3 percent a year—an offset for productivity growth. Individual rates within each basket could rise by no more than 4 or 5 percent a year.

The second basket of services, the 800 services, was deregulated except for Directory 800 in 1993; the third basket was reduced to only analog private-line services in 1991.[29] All remaining services, except for international services, were deregulated in October 1995, when the FCC ruled that AT&T no longer had market power in the domestic interexchange market.[30]

The LEC price caps are more complicated than the AT&T rate cap. First, the interstate access services of the LECs are divided into four categories or baskets: common-line services, traffic-sensitive services, special-access services, and interexchange services. In addition, a revenue-sharing provision has been added to the price caps, thus compromising their effects on efficiency incentives (see chapter 4). Finally, carriers may not raise or lower rates more than 5 percent a year without cost justification.[31] In 1995 the FCC adjusted the productivity offset, increasing it to between 4.0 and 5.3 percent a year, thereby forcing access rates down sharply. The RBOCs are appealing the decision.

DEREGULATION. The FCC was an early champion of deregulating terminal equipment. After succeeding in opening terminal-equipment supply to competition in the late 1960s and 1970s, the FCC then moved to deregulate terminal equipment rentals totally by requiring the detariffing of embedded customer premises equipment in the 1980s.[32] All of the BOCs' customer premises equipment

29. Federal Communications Commission, *Price Cap Performance Review for AT&T*, 8 FCC Rec. No. 19, CC Docket No. 92-134 (July 23, 1993), pp. 6968–86.

30. In the Matter of Motion of AT&T Corp. to be Reclassified as Nondominant Carrier, Order, October 12, 1995.

31. Policy and Rules Concerning Rates for Dominant Carriers, Supplemental Notice of Proposed Rulemaking, 5 FCC Rec. 6786 and Erratum, 5 FCC Rec. 7664 (1990).

32. This means that customers no longer lease their telephone handsets or other

was detariffed at divestiture in 1984, and that of independents was detariffed slowly over the next five years. The FCC has also succeeded in detariffing all *new* inside wiring, but otherwise it continues to regulate interstate services offered by the local exchange carriers.

Under the 1934 Communications Act, the FCC must perform some regulation of all interstate services. For many years the commission has only regulated the rates of dominant carriers, leaving the newer other common carriers (OCCs) essentially to file any tariffs they please. This meant that AT&T's rates were subject to regulatory scrutiny, but MCI's or Sprint's were not. This regulatory forbearance was extended by the FCC in 1985 to full detariffing of all nondominant carriers. In other words, the FCC would no longer require and would not even accept tariff filings from the new competitors. However, AT&T complained that this forbearance violates the clear meaning of Section 203(a) of the 1934 Communications Act, which requires that *every* carrier *shall* file tariffs with the FCC. The Court of Appeals agreed, and the FCC now requires tariffs to be filed by competitive carriers.[33]

In recent years AT&T's share of the interstate interexchange market has fallen to about 60 percent, leading some observers to question whether AT&T was still truly dominant. Some of AT&T's competitors were adamant that AT&T be prohibited from varying its rates freely for fear that AT&T would engage in predation against them. Given the sunk costs of MCI, Sprint, and others, such predation would be unlikely to work; hence the FCC has now ruled that AT&T is now "nondominant" in all domestic long-distance markets.[34] In addition, the FCC has approved greater rate flexibility for AT&T in its customized business tariffs under Tariff 12.[35]

The FCC also continues to regulate the local exchange carriers' interstate access rates through a complex averaging process that

---

terminal equipment at a tariffed rate from their telephone company. Instead, they purchase or lease such equipment from any of a number of vendors, including the telephone company, separately from the purchase of telephone service.

33. *American Telephone and Telegraph Co.* v. *FCC*, No. 92-1053, D.C. Cir., November 13, 1992.

34. See footnote 30.

35. These tariffs allow AT&T to offer a mix of services at rates that are competitive with unregulated carriers. In August 1991, however, the FCC disallowed new Tariff 12 tariffs that include inbound-WATS "800" services.

does not allow the large long-distance carriers to reap the benefits of their heavy traffic.[36] Were these rates to reflect traffic densities, AT&T might obtain substantially lower rates for connecting its subscribers to the LEC switches than do its smaller rivals. Such a result is discouraged, however, by the AT&T decree that mandates equal access for all long-distance carriers.

The "Third Regulator"—The Federal Courts

Since the 1984 AT&T divestiture, the federal District Court for the District of Columbia, the trial court in the *AT&T* case, has assumed an important regulatory role for telephone services. The *AT&T* decree, or MFJ, bars the divested regional Bell operating companies from inter-LATA services and equipment manufacturing.[37] These restrictions were designed to prevent the local Bell operating companies from abusing their bottleneck positions in local telephony by discriminating against unrelated long-distance carriers or equipment suppliers.

Changes in technology and new service offerings have created the need for numerous rulings on the meaning of these restrictions, leading the RBOCs to petition for many exemptions from them. For instance, to offer new information services, the RBOCs may have to provide customers with access to databases across LATA boundaries. If they provide this service directly or as a reseller, they are in violation of the letter of the inter-LATA service line-of-business restriction. Similarly, if an RBOC and a cable company form joint ventures to offer telephone service outside the RBOC's region and if the joint venture straddles LATA boundaries, the service is an inter-LATA service that violates the decree. In addition, cellular companies owned by the RBOCs can only offer long-distance services through resellers in any region of the country even though the RBOC may have no bottleneck local telephone facilities in many of the regions in which its cellular company operates.

The problems created by the AT&T decree are growing with the

36. Carriers are charged a flat rate per minute for access, based on tariffed common-line, switching, and transport elements. Large carriers are not allowed a discount for funneling heavy volume over a given transport link, but they may establish their own transport facilities to the local carrier's switch.

37. The information services restriction was vacated by the U.S. Court of Appeals.

increasing diversity and complexity of modern telecommunications operations. Petitions for exceptions to the decree have backed up before the trial court administering the decree and in the Department of Justice, which must provide recommendations for the court's action. As a result the RBOCs have been pressing for legislation to loosen or abolish the restrictions, and four of the companies have petitioned the court to vacate the decree.[38] The role of the decree and its relevance to the future structure of the telecommunications sector is clearly the most contentious issue in current U.S. telecommunications policy, but there is no consensus on the appropriate policy for preventing anticompetitive behavior by a vertically integrated telephone company with some degree of monopoly power over bottleneck facilities.[39]

Regulatory Authority in Canada

As in the United States, divided jurisdiction has existed in Canada over telecommunications developments. The separation of federal from provincial authority over activities including telecommunications rests primarily with the British North America Act (BNA) of 1867. In the mid-nineteenth century telegraph lines were the means of telecommunications, and they were specifically placed under federal jurisdiction in the BNA Act.[40] The development of new forms of telecommunications (the telephone), however, was not automatically under federal control. The Canadian courts did not accept the argument that all other telecommunications were identical to telegraph communication and therefore under federal jurisdiction. Instead, the courts have had to examine the respective powers of the various levels of government. Federal powers derive from a residual "peace, order and good government" clause and from a general "trade and commerce" power. Provincial governments base their jurisdictional claims for activities under

38. *U.S.* v. *Western Electric Co., Inc. and American Telephone & Telegraph Company, Civil Action No. 82-0192, U.S. District Court for the District of Columbia, Motion of Bell Atlantic Corporation, BellSouth Corporation, NYNEX Corporation, and Southwestern Bell Corporation to Vacate the Decree,* July 6, 1994.

39. See chapter 4 for our discussion of the economics of this issue.

40. British North America Act, 1867 section V 92.10a, p. 149.

constitutional sections that grant provinces authority over "local works and undertakings," "property and civil rights," and "matters of a merely local and private nature within the province."

Jurisdictional Conflicts

An important distinction between Canada and the United States is that Canada has never had a single company that operates across all regions in the manner of AT&T in the United States. Bell Canada, although so authorized in its Federal Charter of 1880, did not successfully step outside the boundaries of Ontario and Quebec;[41] similarly, all other Canadian telephone companies were limited until recently by provincial boundaries.

The Canadian telecommunications system is much more linear than the U.S. system, with 80 percent of the Canadian population living within a hundred miles of the U.S. border. A national telecommunications network in Canada was established by a 1931 agreement between the nine major regional telephone companies in Canada, which cooperated as the TransCanada Telephone System (TCTS).[42] TCTS set out principles for interconnection, established standards for service, and enabled a consistent delivery system from coast to coast. In the absence of a single long-distance company, some settlement procedure had to be developed for passing traffic from one provincial company to another. Various settlement plans were established by the TCTS to enable the sharing of revenues from telephone service across Canada. In addition, an all-Canadian route for the transmission of long-distance telephone calls was built. The resulting Canadian telecommunications network is currently interconnected with the U. S. network through a number of border crossings and to telephone operations outside North America through Teleglobe Canada, which has a monopoly on overseas traffic.

THE FEDERAL JURISDICTION BEFORE 1989. Canadian jurisdictional conflicts had their origin in 1905. Bell Canada, which accounts for 50 percent of telephone revenues in Canada, was incor-

41. Bell Canada owns a controlling interest in Maritime Telephone, but provincial regulators have prevented Bell Canada from exercising control over this company.

42. Later this name was changed to Telecom Canada and then in the 1990s to Stentor.

porated in 1880 by a special statute of the Parliament of Canada. Bell's Charter of Incorporation states that "the works thereunder authorized are hereby declared to be for the general advantage of Canada."[43] In the early 1900s, the city of Toronto attempted unsuccessfully to regulate Bell Canada's local rates by relying on the fact that local calls in Toronto were entirely within the province of Ontario.[44]

In 1906 regulatory jurisdiction over Bell Canada was given to the Board of Railway Commissioners for Canada, whose name was changed in 1938 to the Board of Transport Commissioners for Canada. No major changes in the method of regulation of transport or telephone or telegraph occurred until 1967, when the Canadian Transport Commission was formed to administer the new National Transportation Act. In 1976 the Canadian Radio-Television and Telecommunications Commission (CRTC) was established and given jurisdiction over telephone and telegraph (still controlled under the Railway Act and the National Transportation Act).

With the exception of the British Columbia Telephone Company (BCTel), which also operated under federal incorporation and therefore was subject to federal regulation, all other telephone companies established elsewhere in Canada, either private or publicly owned, were regulated by provincial jurisdictions (see table 2-4). Provincial regulation continued despite the 1905 Privy Council decision on the dispute between Bell Canada and the city of Toronto, which ruled that telephone companies interconnected at the border of the province were in interprovincial trade and therefore subject to federal jurisdiction. The federal government did not attempt to exercise regulatory authority over all telephone companies in Canada. This led to a separation of jurisdictions, with federal jurisdiction over federally incorporated telephone companies and provincial jurisdiction over all other telephone companies. This separation lasted until 1989.

43. Statutes of Canada, May 1882, ch. 9, sec. 4, p. 195.

44. The city's case rested on two points. First, the local aspect of Toronto exchange rates had always been under provincial jurisdiction. Second, in 1880 there were no interprovincial telephone lines in the company. Therefore, it argued, the Federal Charter was meaningless since the company was not a "connecting work" in the sense of interprovincial commerce. Nevertheless, the Privy Council of Great Britain (the Supreme Court of Canada at the time) ruled that Bell Canada was a single undertaking and thus subject to federal authority.

Table 2-4. Major Canadian Telecommunications Carriers and their Regulators, 1988[a]

Carrier	*Regulatory agency*
Bell Canada	Canadian Radio-TV & Telecommunications Commission (CRTC)
British Columbia Telephone Co. (BCTel)	Canadian Radio-TV & Telecommunications Commission (CRTC)
CNCP Telecommunications	Canadian Radio-TV & Telecommunications Commission (CRTC)
Teleglobe Canada	Canadian Radio-TV & Telecommunications Commission (CRTC)
Telesat Canada	Canadian Radio-TV & Telecommunications Commission (CRTC)
NorthwesTel	Canadian Radio-TV & Telecommunications Commission (CRTC)
Terra Nova Telecommunications	Canadian Radio-TV & Telecommunications Commission (CRTC)
AGT	Alberta Public Utilities Board
SaskTel	Responsible to government of Saskatchewan
Manitoba Telephone System	Manitoba Public Utilities Board
New Brunswick Telephone Co. Limited	New Brunswick Public Utilities Board
Maritime Telephone & Telegraph Company Limited	Nova Scotia Public Utilities Board
The Island Telephone Company Limited	Prince Edward Island Public Utilities Board
Edmonton Telephones	City of Edmonton
Northern Telephone Limited	Ontario Telephone Service Commission
Quebec-Telephone	Regie des Services Publiques du Quebec[a]
Telebec Ltee	Regie des Services Publiques du Quebec[a]
Thunder Bay Telephone System	Ontario Telephone Service Commission

Source: Author's survey (compiled by author).
a. In December 1987, the Quebec government introduced a bill aiming at the creation of a new agency to be called the Regie des Telecommunications du Quebec.

Thus the Canadian model was very different from that which evolved in the United States. Whereas in the United States interstate rates were controlled by the FCC and intrastate and local rates by states, in Canada federal jurisdiction was exercised over all aspects of service (including intraprovincial and local rates) for only two of the nine major telephone companies in Canada, Bell

Canada and BCTel. In the United States there was joint supervision of rates with a federal entity regulating interstate rates, and a local regulator regulating local and intrastate rates under a regime of separations of facilities for interstate and intrastate traffic. In Canada there was distinct rather than joint regulation, and the result was that there was essentially no supervision over interprovincial rates.

For many years the CRTC's jurisdiction did not extend outside Ontario, Quebec, and British Columbia. As a result, it did not regulate rates for calls between these provinces and the Maritime or Prairie provinces. The CRTC first began to show an interest in regulating these interprovincial rates in 1978 in response to applications for interprovincial rate increases by Bell Canada and BCTel. The CRTC's interest in these interprovincial rates was triggered by Canadian Pacific's application to compete in private-line services two years earlier, an application later joined by the Canadian National Railway Company resulting in the formation of CNCP. This led the CRTC into the process of controlling interprovincial rates, but only insofar as the rates for Bell or BCTel could be judged excessive. An accounting firm, Peat Marwick, was hired to examine and compare the system used for settlements among members of TCTS, the RSP (Revenue Settlement Plan), adopted in 1977.[45]

The RSP allocated costs, variable and fixed, to the provision of interprovincial traffic. Carriers settled for the difference between the revenues they generated and the costs that were attributable to interprovincial service. These contributions from long-distance services were very large. For instance, in 1978 Bell Canada paid 24 percent of its total originating interprovincial revenues as settlements to other companies for completing its calls. In 1981 the CRTC decided that Bell Canada was contributing in excess of $40 million to the RSP pool and ordered its interprovincial rates cut.[46]

There are also jurisdictional conflicts at the federal level over the division of authority between day-to-day analysis of rates and

45. Peat Marwick and Co., *Phase Two Report, A Review of TransCanada Telephone System Originated Revenue, Settled Revenues and Settlement Account Balances,* November 1979, figure VI-2.

46. CRTC Telecom Decision 81-13, July 7, 1981.

long-term policy. The CRTC has regulatory authority extending only to the analysis of rates and the revenue requirement (including capital construction). Jurisdiction over crucial issues such as microwave licenses is vested in the federal Department of Communications, which also sets policy goals. In addition, the federal cabinet hears appeals from CRTC decisions on all matters involving both rate issues and the merits or demerits of competition.[47]

Federal communications regulation evolved through a panoply of different laws: the Railway Act already discussed, the Radio Act, the Telegraph Act, the Broadcasting Act, the Teleglobe Canada Act, the Telesat Canada Act, the Department of Communications Act, and the Canadian Radio Television Telecommunications Act, among others.[48] Until the recent passage of a new Telecommunications Act (1993), the streamlining of regulatory authority in one consistent act was successfully blocked in the Parliament.

PROVINCIAL REGULATORY AUTHORITY. Provincial regulation has varied considerably across the various provinces. A majority of provincial regulatory boards regulate other services as well. For example, in Nova Scotia, Maritime Telephone and Telegraph is regulated by the Nova Scotia Public Utilities Board, which regulates electric utilities, bridge tolls, gasoline prices, the hours of service stations, and northern carrier licenses as well as telecommunications. Saskatchewan Telecommunications (SaskTel) is a crown corporation controlled by directors appointed by the government and reporting to its holding company, the Crown Investments Corporation, and to a select standing committee of the legislature. A statutory regulatory agency oversees SaskTel.

It was only in the early 1970s that the overlap or conflict between federal and provincial jurisdiction became a major public issue. A number of provinces, particularly Quebec, Manitoba, Saskatchewan, and Alberta, became concerned that the federal government

47. The Federal Cabinet cannot provide directives to the CRTC, whereas several provincial jurisdictions did provide for this ex ante policy input. The Federal Cabinet's ability to rewrite or rescind CRTC decisions has been controversial since the process at Cabinet level is neither transparent nor necessarily "fair." No province allowed appeals to Cabinet; appeals had to be routed through the Courts.

48. R. Brian Woodrow and others, *Conflict over Communications Policy: A Study of Federal-Provincial Relations and Public Policy* (Montreal: C. D. Howe Institute, 1980).

was attempting to establish domestic communications policy without any provincial involvement.[40] These fears had begun in 1966 when the federal government outlined its policy for Telesat Canada as a monopoly carrier for satellite communications and did so without any involvement from provincial governments.

In April 1973 the federal government offered its proposals for a communications policy for Canada, setting out a national mandate and an expansion of the federal jurisdiction and regulatory authority. Beginning in the mid-1970s, the provinces coordinated their views. In 1975 they submitted a joint provincial statement, which rejected federal controls and suggested devolution of most regulatory authority to the provinces. By the late 1970s a fundamental difference existed between the federal and provincial views of the jurisdiction over telecommunications, with the federal government pushing for federal supremacy and the provinces requesting full devolution and an absence of federal jurisdiction except for interprovincial rates and issues of national concern. The provincial principles would have led to the breakup of Bell Canada and a division of regulatory authority between the provinces of Ontario and Quebec. The unique role of the province of Quebec within the Canadian confederation and Quebec's views of the importance of culture and the role of communications within culture have been major forces within this movement. Provincial cries for increased authority continued to mount as competitive pressures and threats appeared.

In the late 1970s two issues of competition became important, confirming the provinces in their view that they should have sole authority over most telecommunications matters. In 1976 Canadian Pacific (later CNCP) first sought permission for interconnection between its long-distance facilities and Bell's local system for the provision of competitive private-line service. In 1978 Challenge Communications challenged Bell Canada's exclusive provision of terminal equipment, resulting in a liberalization by the CRTC of terminal attachment practices.

Federal-provincial conferences were called for communications ministers to attempt to sort out these new issues. Jurisdiction over communications became embroiled within the broader and more

49. Ibid.

significant question of Canadian constitutional reform. The issue of regulation on some joint federal/provincial model (such as the U.S. model) assumed greater importance.[50] Federal Canadian telecommunication legislation was delayed, and federal decisions for 1980s services and facilities continued to be based on interpretation of the 1905 Railway Act. In February 1986, a committee of ministers (provincial and federal) was established to provide a framework for the establishment of a new telecommunications policy. It took two years to agree on six policy principles—"a unique Canadian approach"; universal access; "international competitiveness of Canadian industry"; technological progress; "fair and balanced regional development"; and the role of government. Unfortunately, these six broad goals could be applied to any Canadian policy; a fundamental reform of telecommunications policy, requiring much more specific proposals, remained to be articulated by the federal government, which finally exercised jurisdiction when the Supreme Court recommended that it do so in 1989.

In 1989 the Supreme Court issued a decision that clearly established the federal government's jurisdiction in regulating all Canadian telecommunications except for the provincially owned utilities.[51] This decision launched a process of revising the antiquated Railway Act, which had been the basis of federal regulation for most of the century. The Committee of Ministers and the pressures for provincial control initially suggested a regulatory model involving provincial decisionmaking.[52] However, the 1993 Telecommunications Act provides for no such provincial input.

Regulating Competition

In 1975 there was little competition in any Canadian telecommunications activity, not even in terminal equipment. Regulated telephone companies had monopolies over terminal equipment, inside wiring, and outside wiring. There was no facilities-based

50. Hudson N. Janisch and Richard Schultz, "Federalism's Turn: Telecommunications and Canada's Global Competitiveness," *Canadian Business Law Journal,* vol. 18 (August 1991), pp. 161–87.

51. *IBEW* v. *Alberta Government Telephones,* Supreme Court of Canada, August 14, 1989.

52. See Janisch and Schultz, "Federalism's Turn."

competition for long-distance and local services. Some believe that Canada generally lags behind the United States by a decade; in telecommunications this proves to be an understatement. In the United States the first glimmerings of competition in the long-distance market could be seen in the 1950s with the *Above 890* decision and the liberalization of the provision of private-line services by specialized common carriers. In Canada the Department of Communications did not approve private microwave licenses for specialized common carriers.

SPECTRUM. The allocation of the radio spectrum is clearly a federal responsibility in Canada since radio waves cross-provincial boundaries. As a result, the Department of Communications was able to restrain the development of competition both federally and provincially.[53] This restraint, of course, was approved by provincial regulators, who have shown an antipathy to competition over the years. While CNCP Telecommunications was allowed to offer competitive private-line service in 1979, CNCP could not offer this service outside Ontario, Quebec, and British Columbia, the only provinces then under federal regulation.

TERMINAL EQUIPMENT. On November 23, 1982, the CRTC reaffirmed its interim decision for a liberalized terminal attachment policy, stating that single-line residential subscribers could own all their telephone sets including the main set as long as the equipment met the technical requirements established by Communications Canada's terminal attachments program.[54] Inside wiring, however, would remain the property and responsibility of the telephone companies except for multiline business customers, who would also own the associated inside wiring when they owned their own terminal equipment.

LONG-DISTANCE SERVICES. The deregulation of terminal equipment occurred nearly a decade after the U.S. decision to liberalize. The CRTC approved CNCP's applications to offer private-line services in 1979, but it stated that MTS and WATS, the switched

53. Radio common carriers have been allowed to operate. "Most radio common carriers are community-oriented enterprises serving a relatively restricted geographical area." Steven Globerman, "Deregulations of Telecommunications: An Assessment," in Walter Block and Geroge Lermer, eds., *Breaking the Shackles: Deregulating Canadian Industry* (Fraser Institute, 1991), p. 89.

54. CRTC Telecom Decision 79-11 (May 17, 1979).

long-distance services, were to be protected from direct competition.[55] In that decision, the CRTC required a contribution charge from CNCP, an access charge payable to the telephone companies for interconnection that exceeded the incremental costs to these companies of providing such connections. This contribution is mandated to maintain low local rates. In this decision the CRTC established that these contribution charges would be a guiding principle for regulatory policy.

In October 1983 CNCP Telecommunications applied to the CRTC for permission to offer public switched service for both MTS and WATS in Bell Canada's territory. The 1985 CRTC decision rejected the application. While the CRTC found that interconnection and competition would lead to consumer benefits, the commission was not convinced that CNCP would actually be able to operate profitably given the contribution payments that were required. This concern over the financial viability of an applicant seems odd, given the long history of CNCP in Canada (CNCP constructed the first coast-to-coast microwave system) and its deep pockets. More important was the commission's concern that local rates might have to increase as a result of telephone companies' loss of revenues in long-distance service.

In 1989 Canadian National's interest in CNCP was sold to Canadian Pacific. Rogers Communications, the largest cable operator in Canada and also one of the two holders of cellular telephone licenses, purchased 29.5 percent of the new company, renamed Unitel. Unitel reapplied in 1990 for the right to offer switched MTS and WATS service across Canada. After a long hearing, the CRTC approved the application, essentially reversing its 1985 decision.[56] The pressure to liberalize telecommunications services had become too great, and the successful transition to greater competition in long-distance services in the United States and the United Kingdom could no longer be ignored. A new direction in Canadian telecommunications was now politically acceptable, a conclusion reinforced by Parliament's decision to extend far greater authority to the CRTC in the 1993 act.

55. Specifically, the CRTC approved a petition from CNCP requesting interconnection with Bell Canada's local circuits for CNCP's provision of dedicated private-line service in Decision 79-11.

56. CRTC Decision 92-78, Telecom Public Notice (December 16, 1992).

Unitel was able to apply for interconnection of its network with all local telephone carriers across Canada because of the crucial Supreme Court decision of 1989 that shifted regulatory jurisdiction to the CRTC. After the CRTC decision of 1979 allowing private-line interconnection with Bell Canada and BC, CNCP began negotiations with Alberta Government Telephones (AGT) for similar interconnection rights, but these negotiations failed. When AGT applied to the Federal Court of Appeals for an order quashing the CRTC's right to order interconnection for competition outside Bell Canada and British Columbia Telephone, the matter ended up in the Supreme Court, where a 1989 decision extended federal regulation to all private telecommunications services. The three government-owned telecommunications companies in the prairie provinces were not, however, subject to this decision because of Crown immunity. The federal government moved in 1989 to introduce legislation to put the three provincial telephone companies under federal jurisdiction, and simultaneously AGT was privatized.[57]

The 1993 Act revolutionized Canadian telecommunications regulation by requiring:

—Federal regulation of all telecommunications in Canada except over provincially owned utilities for a period of five years;[58]

—CRTC authority to forbear from regulation where competition is likely to be of sufficient force to protect consumers;

—Distinctions between regulated and exempt services. Telecommunications common carriers and entities owning telecommunications facilities and transmission facilities are subject to regulation. Any transmission apparatus that performs switching, input-output, or processing of information or control of the speech, protocol, or format of transmission is exempt from regulation;

—A more open process for Cabinet appeals, though it continues to allow all decisions of the CRTC to be appealed to the Cabinet, which may vary or rescind decisions;

57. Hudson N. Janisch, "At Last! A New Canadian Telecommunications Act," *Telecommunications Policy*, vol. 17, no. 9 (December 1993), pp. 691–98.

58. The 1993 Telecommunications Act provided for the CRTC to assume regulatory authority over the two telephone companies owned by the provincial governments of Manitoba and Saskatchewan in 1993 and 1998, respectively.

—CRTC authority to exempt any class of carriers from regulation, but only after a public hearing.

INCREASING LOCAL COMPETITION. Having admitted entry into interprovincial services, the CRTC next turned to the issue of a regulatory framework for Canadian telecommunications in the environment of competitive and monopoly services. After long hearings, the commission issued a sweeping set of new approaches in September 1994.[59] It embraced a policy of substituting competition for monopoly wherever possible, including local access. To accomplish this, the commission ordered telephone companies to unbundle their network services and to offer collocation and reciprocal access to all competitive carriers. Perhaps the most startling aspect of the CRTC's decision was its announcement that it would seek to rebalance local and long-distance rates by requiring three annual local-access rate increases of $2 (Canadian) per month in 1995, 1996, and 1997. These increases would allow the local companies to reduce carrier access charges for long-distance services, thereby reducing the incentive for carriers to seek uneconomic bypass of the network.

After the three-year phase-in of access charge reductions, the CRTC would begin to regulate through rate caps similar to those used by the FCC in the United States. In the interim, rate-of-return regulation will continue with revenue sharing and different allowed rate-of-return ranges for interexchange and "utility" (local) services. In addition, the CRTC required that all services offered by vertically integrated carriers (such as Bell Canada) meet an imputation test that requires rates to be set at or above the carrier's charges to competitors for its own network elements plus its incremental cost of additional inputs. Thus, for example, Bell Canada would have to charge long-distance customers rates that were at least as great as the sum of its carrier access charge plus its own incremental transmission, marketing, and administrative costs. The commission explicitly rejected any other form of rate floors.

The CRTC also opened the door to video competition by allowing telephone companies to offer video dial tone, a common-carrier service to be programmed by third parties. In a sharp contrast with U.S. policy, however, the CRTC rejected vertical divestiture and

59. CRTC Telecom Decision 94-19.

structural separation as approaches for dealing with the potential for anticompetitive abuses by integrated telephone companies. The commission concluded, "Divestiture could be damaging to the competitiveness of Canada in global markets and could dampen the emergence of integrated services evolving from the convergence of the communications, information, computing and entertainment industries."[60]

Less than three months after the CRTC announced its new regulatory framework, the government announced a suspension and reevaluation of the rate rebalancing conducted through the three annual $2 per month local-access rate increases. Because this policy was designed to reduce carrier access rates, it was immediately opposed by Unitel, the new long-distance carrier. In admitting Unitel in 1992, the CRTC had provided the company with a subsidy in the form of sharply reduced carrier access rates, reductions that were to be phased out over five years. When the CRTC announced its 1994 decision to reduce all carrier access rates through rate rebalancing, it implicitly reduced the start-up subsidy for Unitel. When Unitel began to encounter financial difficulties in 1994, it added its voice to the populist calls for a reversal of the commission's decision to rebalance rates.[61]

ENHANCED SERVICES, RESALE, AND SHARING. The CRTC currently defines carriers as Type I (terrestrial facilities providers) and Type II carriers. The latter are prohibited from owning and operating basic transmission facilities but can engage in resale and sharing and can offer enhanced services. Federal supervision at times appears to be expanding, not declining, but not as a result of CRTC pressure. For example, the federal court, on an appeal from the Telecommunications Workers Union, held that resellers of long-distance service are telephone companies under the Railway Act. Hence these resellers must file tariffs. The CRTC had attempted to forbear from such regulation.

The time lags relative to the United States have also been substantial in other policy areas. In 1984 the CRTC defined enhanced

60. CRTC Telecom Decision CRTC 94-19, section II.

61. In 1995, the CRTC reaffirmed its policy of rate rebalancing, although the third $2 per month local rate increase was to be decided in the future. As this book goes to press, however, this issue is still being contested by the parties.

services, distinguishing them from basic services.[62] Basic service was defined as offering a pure transmission capability over a communications path that is virtually transparent. Message-telephone services (MTS) and WATS services were defined as basic services. The remaining enhanced services were allowed to grow with limited regulation and liberalized entry. In 1987, the CRTC allowed resale and sharing of services other than public switched telephone services (for example, MTS and WATS).[63] Resale and sharing were to be allowed for private-line services as defined in the 1979 CNCP application and for enhanced service providers. To prevent bypass of MTS and WATS, two restrictions were imposed: the circuits would have to be dedicated to a sharing group; and interconnection was to be provided by the telephone company for any member of the sharing group having interconnected voice requirements. "Hence, the *Resale and Sharing* decision was predominantly focused on the restrictions required to prevent resellers from offering MTS/WATS-like telephone services and containing the growth of competitive service offerings to private lines."[64]

In 1987 a Canadian firm, Call-Net, attempted to offer a service called CDAR to erode the distinction between private-line and switched services in much the same fashion as MCI used its Execunet service to break into the switched MTS/WATS market in 1975. The Call-Net service was basically switched MTS with a sophisticated billing and accounting function, CDAR. The CRTC forbade Call-Net from offering its service, much as the FCC had attempted to block MCI from offering Execunet. At first Bell Canada refused to supply interexchange facilities to Call-Net on the grounds that the service was "basic," even though it was up to the CRTC, not the monopoly facilities provider, to define whether services were basic or enhanced. The CRTC held a hearing and in its decision stated that while CDAR itself and its equipment was enhanced service, CDAR was separable from the transmission path and the primary function of the service package was basic service.[65] As a re-

62. CRTC Telecom Decision 84-18 (July 12, 1984).

63. CRTC Telecom Decision 87-2 (February 12, 1987).

64. Terry Dawn Hancock, "Regulated Competition: An Analysis of Resale and Sharing Competition in Telecommunications," LLM thesis, Faculty of Law, University of Toronto, 1991, p. 42.

65. CRTC Telecom Decision 87-5 (May 22, 1987).

sult, the CRTC ordered Call-Net to desist from offering the service and to disconnect its transmission facilities (from Bell Canada) within thirty days.

On the day it was to be disconnected, Call-Net appealed to the Canadian Cabinet, which basically stayed the execution for some 300 days. A compromise failed, and Call-Net appealed to the federal court. Unlike the U.S. courts, however, the Canadian federal courts denied the application, and leave to appeal to the Supreme Court of Canada was denied. Thus an attempt to enter MTS/WATS service through the back door in Canada was not allowed.

In four decisions the CRTC expanded resale and sharing of non-WATS, non-MTS services, always carefully attempting to prevent erosion of basic revenues.[66] Decisions 92-11 and 92-12, allowing interexchange competition, include new rules for resellers. CRTC 92-11 regulates resellers by requiring filings and CRTC approval of tariffs. CRTC 92-12 set the 1993 level of access payments by resellers at 65 percent of the contribution paid by the established (facilities-based) carriers, a rate that was to increase to 85 percent in 1997.

CELLULAR AND PERSONAL COMMUNICATIONS NETWORKS. Canada's cellular policy was announced by Communications Canada in October 1982. In cellular radio there is not the same jurisdictional fight over regulatory authority as there was with land-line systems because the regulation of the radio spectrum is clearly the jurisdiction of the federal government. The announced Canadian policy was similar to the policy in the United States, allowing a maximum of two systems in each service area—one for the local telephone monopoly, and the other for another system provider. Compatibility with each system and with the United States was ensured. Seven industry groups and nine telephone companies submitted license applications in February 1983; ten months later the local land-line telephone companies were given licenses, as was Cantel, Inc., a subsidiary of Rogers Cable. Thus in Canada, unlike the United States, the nonwire-line carrier is licensed nationally. The cellular operations of land-line companies are marketed separately, but these carriers have an association, CellNet

66. CRTC Telecom Decisions 85-19 (August 29, 1985); 87-2 (February 12, 1987); 90-3 (March 1, 1990); 90-19 (September 4, 1990).

Canada. In 1994, Mobility Canada had approximately 58 percent of the Canadian cellular market.[67] These carriers are not regulated.

In December 1992 the CRTC awarded four national licenses to operate PCN services (digital and in the 944–48 MHz range).[68] Cantel and Mobility Canada (a subsidiary of BCE Mobile Communications) were awarded licenses. Thus two of the four PCN licenses are held by cellular providers. There have been limited trials, but otherwise no movement to a PCN system has been announced. Cantel stated in its 1993 annual report that it had "reassessed its PCN business plan given the terms and conditions" imposed. Several of the licensees have had discussions about joint construction of a network. However, given the U.S. decision to auction spectrum in the 1.8 GHz band for PCN in June 1994, the Canadian government announced on June 15, 1995, that six blocks of spectrum in the 2 GHz band (three 30 MHz and three 10 MHz blocks) would be licensed in 1995.[69]

In 1988 the Canadian government decided that one satellite for mobile services would be launched; it would be owned by Telesat Canada, the monopoly provider of domestic and Canadian-U.S. satellite services. In May 1993 the Canadian government reacted to the emergence of worldwide satellite systems such as Motorola's IRIDIUM Project by proposing that the licensees for the radio apparatus and ground stations would be limited to a maximum of 20 percent foreign ownership (this is consistent with ownership rules for telecom generally). In addition, the applicant would have to prove that the services offered provide a demonstrable benefit to Canadians and that universal services would not be put at risk.[70] The Canadian government at first, in effect, prevented Hughes DirecTV (popularly known as "Death Star") from offering its services to Canadian consumers[71] by exempting DBS services carried

67. Mobility Canada, "A Spectrum of Choices for Canadians" (1995).

68. Canada uses the acronym PCN (Personal Communications Networks) to describe the Personal Communications Services (PCS) described in chapter 1.

69. Industry Canada, Press Release, June 15, 1995.

70. Communications Canada, "Proposed Policy for the Provision of Mobile Satellite Services vs. Regional and Global Satellite Systems in the Canadian Market," Ottawa, May 1993.

71. CRTC, Broadcast Public Notice 1993-74, p. 8 (June 3, 1993). Hughes DirecTV is the U.S. Direct Broadcast Satellite Service that began operation in mid-1994.

on Canadian satellites, which left one service—ExpressVu, a consortium of Bell Canada Enterprise, Cancom, Tee-Com, and WIC (Western International Communications). The Canadian government intervened,[72] directing the CRTC to reexamine its policy so as not to result in a single DBS provider. The CRTC announced hearings beginning October 30, 1995, to license several DBS systems.

LOCAL AREA NETWORKS (LANS) AND CABLE TELEVISION. Under the CRTC, noncarrier-provided exchanges could interconnect with the facilities of the federally regulated carriers (Bell, BCTel, and Unitel).[73] Thus cable companies could provide interconnection for LANs. As in the United States, telephone companies have not been allowed to carry broadcast signals.[74] Thus the competition, at this point, is only in one direction: cable companies are positioning for some form of local service as CAPs, but telephone companies cannot compete with cable television. However, the CRTC has begun to respond to the government's request for guidance in designing a policy for the information superhighway.[75] The ground rules for telephone company entry into video services will emerge from this process.

In 1992 the CRTC ruled that the CAP facilities of Rogers Cable qualified as a telephone facility under the Railway Act, and thus Rogers had to file tariffs for these services.[76] Until recently, Rogers and representatives of the Canadian Cable Television Association have denied that they intend to provide local service in competition with the telephone companies, but they now appear to be changing their minds.[77]

72. Orders-in-Council P.C. 1995-1105 and 1995-1106.

73. CRTC Telecom decision 77-6.

74. The Bell Canada Special Act prevents Bell Canada from holding a broadcast license. Other telephone companies are not explicitly barred from owning such a license, but none has been granted to a telephone company.

75. Public Notice CRTC 1994-130 in response to Order in Council PC 1994-1689.

76. CRTC Telecom Decision 92-10 (June 11, 1992).

77. "Rogers Aims to Pull Plug on Rival," *Globe and Mail*, May 30, 1994, p. B1. In 1995 the CCTA argued before the CRTC that telephone companies should not be allowed to offer video services until the cable companies were able to compete in the local telephony market and had completed a major network modernization program.

Regulating U.S.-Canadian Telephone Services

Telecom traffic and revenue between the United States and Canada is substantial. In 1993 calls to the United States composed two-thirds of all international calls emanating from Canada, amounting to $700 million. Calls from the United States to Canada were more numerous. The United States has a trade deficit with Canada in international calls, reflecting the far greater volume of U.S.-Canada relative to Canada-U.S. calls even when the higher rates from the United States to Canada are taken into account.

The Canadian and U.S. telecom systems are completely integrated and connect at many border points. As a result of this interconnection, U.S.-Canada traffic has not been treated or regulated in ways similar to Canadian international and U.S. international telecommunications traffic. Teleglobe Canada is a monopoly service provider, interconnecting Canada and all foreign jurisdictions except the United States and Mexico. The Canadian telephone companies are the providers of Canada-United States traffic. Before MCI's entry, AT&T had a monopoly on long-distance switched telecommunications overseas or to North America. This extended to calls to Canada. All international calls, whether U.S.-Canada or U.S.-Japan, require each country to set prices for outgoing calls (called "settlement rates") and to negotiate a means of paying the other country for the use of the facilities in terminating the call.[78] Countries' carriers negotiate bilateral "accounting rates"; one-half the accounting rate is the amount received by the country (carrier) on the incoming call as payment for its costs in terminating that call. The current accounting rates for Canada-U.S. traffic are $0.26–0.28 peak ($0.22–0.24 off-peak).[79] Thus for a one-minute call from the United States to Canada, AT&T (or MCI) pays its Canadian correspondent $0.14 (at peak, $0.12 off peak) for the costs in Canada of terminating that call.

The FCC has attempted to allow resale of international service

78. See Leland L. Johnson, *Competition, Pricing, and Regulatory Policy in the International Telephone Industry* (RAND Corporation, July 1989); and Organization for Economic Cooperation and Development, Information Computer Communications Policy 34, *International Telecommunications Tariffs: Charging Practices and Procedures* (Paris: OECD, 1994).

79. See FCC Common Carrier Bureau, International Policy Division, *Consolidated Accounting Rates of the United States* (November 1, 1994), p. 1.

to Canada, but AT&T has resisted, arguing that Canada's regulatory system has not opened up its market sufficiently to allow equivalent terms of access to U.S. carriers.[80] Chapter 5 shows that rates from the United States to Canada are considerably higher than those from Canada to the United States. This difference may explain AT&T's desire to resist the opening of the market to resale.

Conclusion

It is quite clear that the United States has moved much further toward liberalization of entry and regulatory reform in communications than has Canada. Canada has only recently allowed entry into switched long-distance services, but it has not liberalized its spectrum policy to allow a proliferation of radio-based telephone technologies.

In both Canada and the United States, regulators have resisted liberalization of entry at various times because of their desire to perpetuate a system of subsidies purportedly designed to induce universal service. In the United States, however, the advocates of competition have been much more successful, and U.S. regulators are thus forced now to consider how to adjust regulation to the mixture of competition and monopoly that has developed. In subsequent chapters, we gauge their success in this respect.

In Canada, on the other hand, market liberalization has been a more recent phenomenon. As the result of a 1989 court decision and the 1993 Telecommunications Act, the CRTC now has jurisdiction over most Canadian telecommunications rates.[81] A sweeping set of procompetitive regulatory policies was announced by the CRTC in 1992 and 1994, but the government has delayed some of them. Whether regulatory reform and market liberalization remain on course remains to be seen.

80. See FCC, *In re Applications of fONOROLA Corporation,* File No. ITC-91-103, November 4, 1992; and EMI Communications Corporation File No. ITC 91-050, 7 FCC Rec. No. 23.

81. *IBEW* v. *Alberta Government Telephones,* Supreme Court of Canada, August 14, 1989.

Chapter Three

Telephone Rates and the Cost Structure of Regulated Companies

EVEN THOUGH CURRENT discussions of telecommunications policy focus on the evolution of the network to an interactive, switched broadband configuration, telephone companies are still principally in the business of delivering traditional voice or data communications through copper-wire subscriber lines. As a result, regulators must still grapple with traditional issues involving local and long-distance rates. This becomes more difficult in an environment of competitive entry and evolving new services. But if regulators do not allow carriers to establish an economically efficient rate structure, they will be forced to impede entry and to limit the carriers' flexibility in developing new technologies and offering new services. Thus efficient pricing of existing services becomes the sine qua non of liberalization.

Although Canadian and U.S. telephone sectors have been liberalized substantially, they are still regulated. Regulators dictate not only the maximum rate level but the structure of rates. As a result, telephone rates are far from economic costs, and market signals for entry, investment, and output growth are significantly distorted.

This chapter explores the cost structure of the regulated firms with the goal of identifying the departure of rates from costs. The focus is ordinary switched voice services and access. These services account for the bulk of industry revenues and are the basis for continued government regulation. As modern telecommunications networks become vastly more complicated, any attempts to regulate rates or service offerings will become much more difficult, if not impossible. But if regulators cannot come close to establishing a climate in which plain old telephone service is priced efficiently, there cannot be much hope that regulation of interactive video services, movies on demand, and information services will be efficient.

The Cost Structure for Local Exchange Companies

There are at least three approaches to estimating the cost structure of the telephone industry. First, econometric techniques can be applied to the estimation of cost or production functions from cross-sectional or time-series data. Second, estimates may be derived from accounting data if one is willing to use engineering or other a priori methods for allocating joint and common costs. Finally, engineering models may be built from current knowledge about equipment design, service requirements, and input costs.

Each of these methods has something to recommend it, but none is perfect. Econometric techniques, favored by economists, are subject to substantial problems of measuring capital stocks (or services) and technical change. Accounting estimates suffer from the vagaries of historical costs, arbitrary allocations of common or joint costs, and arbitrary conventions regarding depreciation of capital stocks. And although engineering models would appear to avoid most of these problems, they are often unable to capture all of the subtleties of today's complex technologies and the routine administrative, billing, and managerial costs.

ECONOMETRIC MODELS. Most of the econometric estimates of telecommunications costs or production functions are derived from time-series models. These econometric models are not sufficiently disaggregated (nor accurate) to provide insights into the costs of delivering business access, residential access, private lines, or switched use.[1]

A recent set of studies uses econometric methods to estimate the determinants of telephone company costs from accounting measures of costs reported to regulatory authorities. Thus this

1. A useful survey of the econometric evidence on economies of scale and scope in telecommunications is in Leonard Waverman, "U.S. Interexchange Competition," in Robert W. Crandall and Kenneth Flamm, eds., *Changing the Rules: Technological Change, International Competition, and Regulation in Communications* (Brookings, 1987), pp. 62–113.

literature represents a hybrid of econometric and accounting methodologies.

An example of a straightforward econometric methodology is the study by Perl and Falk of access and usage costs for thirty-nine local-exchange companies.[2] They simply estimate a linear model of access, local use, and long-distance use to obtain estimates of the marginal cost of each. Their results suggest a large disparity between the marginal cost of an access line, $25 per month (1984–87 dollars), and the typical residential rates prevailing at the time. They also estimate that the local companies' costs of originating or terminating long-distance calls was no more than $0.02 a minute on electronic switches. Moreover, they found that pricing access and use at estimated marginal costs would result in revenues that are between 20 and 30 percent less than average costs. Rohlfs undertook a similar analysis of the costs of the Chesapeake & Potomac (C&P) local exchange operations in Washington, D.C., estimating the costs of various services from time-series regressions on components of C&P's costs.[3] His results showed that the marginal cost of business access in 1984 dollars was $130 per year while the marginal cost of residential access was $600 a year. Rohlfs explains that this difference in costs derived largely from differences in loop length, but he does not provide data on the distribution of loop lengths. Rohlfs concludes that business access was priced very close to cost in 1983, but that residential access rates recovered only $103 of the $600 in annual costs.

Recent articles by Shin and Ying and by Palmer use cross-sectional data from local exchange companies to estimate much more flexible translog cost functions, but even these estimates are of limited use in determining the proximity of telephone rates to economic costs.[4] Shin and Ying estimate a translog cost model from 1976–83 FCC accounting data on carrier costs. Their results sug-

2. Lewis J. Perl and Jonathan Falk, "The Use of Econometric Analysis in Estimating Marginal Cost," paper presented at the Bellcore and Bell Canada Industry Forum, National Economic Research Associates, Inc., April 6, 1989.

3. Jeffrey H. Rohlfs, *Marginal Costs of Telephone Services in Washington, D.C.*, Shooshan & Jackson, Inc., October 1983.

4. Richard T. Shin and John S. Ying, "Unnatural Monopolies in Local Telephone," *RAND Journal*, vol. 23 (Summer 1992), pp. 171–83; Karen Palmer, "A Test for Cross-Subsidies in Local Telephone Rates: Do Business Customers Subsidize Residential Customers?" *RAND Journal*, vol. 23 (Autumn 1992), pp. 415–31.

gest that there are very limited economies of scale in local exchange service, and increasing costs of access with respect to the average length of the loop connecting subscribers.[5] Given their accounting measure of the price of capital and their approximation of the capital stock from reported book values, these results must be viewed with caution. Shin and Ying's translog model yields an estimate of the elasticity of costs with respect to loop length of 0.14. More recently, Shin and Ward have used the Shin-Ying methodology to estimate incremental cost of access in the 1976–87 period for a Federal Trade Commission staff filing before the FCC.[6] They estimate that the marginal cost of subscriber access in 1987 averaged $33 a month in 1983 dollars. Their model would predict that a trebling of average loop length—from 6,500 feet to 19,500 feet—would increase the marginal cost of access to only $38.50. The FTC staff study also estimates the marginal cost of local calls to be $0.02 and the marginal cost of long-distance calls to be $0.12.

Palmer also estimates a translog cost function for switching costs from accounting data, but her capital-stock measure is constructed from investment flows and economic depreciation rates using the perpetual inventory method. The local-loop costs are not estimated econometrically, but rather are simulated from an engineering model developed by New England Telephone, the company that provides her confidential data. Her model focuses on the cost of access and use in the local exchange because she is interested in testing for cross subsidies between business and residential customers. Her results confirm extensive subsidies from business to residential access.[7]

ACCOUNTING DATA. It is also possible to use accounting data directly to estimate the costs of access and use, assigning costs to

5. Specifically, Shin and Ying find that there are slight economies of scale, but that a monopoly of local exchange services would produce at higher costs than two firms for about two-thirds of the output combinations they examined.

6. Comment of the Staff of the Bureau of Economics in CC Docket No. 91-141, transport phases 1 and 2, *Expanded Interconnection with Local Telephone Company Facilities,* March 5, 1993.

7. Specifically, she concludes that in virtually every central office in her study, business access costs are above stand-alone costs. In most (54 to 64 percent) of these offices, residential rates are below incremental costs. Thus, her results support the strict requirements for cross subsidies laid down by Gerald Faulhaber, "Cross-Subsidization: Pricing in Public Enterprises," *American Economic Review* (December 1975).

various categories of service. Unfortunately, such analyses are likely to be quite misleading because they fail to account for inflation in asset prices, require arbitrary separations of fixed and common costs among service categories, and do not adequately reflect technological change.

The Federal Communications Commission publishes estimates of non-traffic-sensitive costs for U.S. local exchange carriers based on its accounting rules. These estimates are currently prepared by the National Exchange Carriers Association. The resulting NTS revenue requirements are approximations of the cost of providing access to the local loop. In 1990 these costs averaged $231 per line for all reporting local exchange carriers, but there is no distinction between business and residential lines, nor between rural and urban lines, in these tabulations.[8]

ENGINEERING MODELS. Given the difficulties in estimating econometrically the costs of local-exchange carriers and the obvious problems of accounting studies, a number of engineering-economic models of telephone costs have begun to appear in the literature. As mentioned above, Palmer's study derives local-loop costs from an engineering model developed by New England Telephone. Gabel and Kennet developed an engineering model of analog and digital local networks.[9] This model was then used to generate several hundred observations on the basis of which the authors estimated a translog cost model. Their conclusions are that the economies of scope in telephone service are limited to markets of low density and that, therefore, local telephone service is not a natural monopoly. They also conclude that digital networks increase access costs but reduce use costs.

The New England Telephone (NET) model of loop costs provides some corroborative estimates of the increasing cost of access as one moves from dense to less dense markets. In 1986, NET estimated the costs of telephone service across its primary calling

8. Our analysis of the welfare effects of regulatory rate distortions is based on 1990 data. The most recent datum on average NTS per line is for 1993: $243. National Exchange Carrier Association, Universal Service Fund, as reported in FCC, Common Carrier Bureau, Industry Studies Division, *Monitoring Report*, May 1995, p. 82.

9. David Gabel and Mark Kennet, "Estimating the Cost Structure of the Local Telephone Exchange Network," National Regulatory Research Institute, Ohio State University, October 1991.

areas in Massachusetts.[10] This study showed that the embedded cost of local residential service in the smallest primary calling areas was 2.4 times the cost of residential loops in the most urbanized locations.[11] These estimates do not include the investment in distribution lines, but the cost of distribution lines also undoubtedly increases with declining wire-center size.

Another engineering-economic model of local telephony has been developed by Mitchell, using data provided by Pacific Bell and GTE of California.[12] He develops the average incremental costs of access and use for three prototypical communities in California—a small urban community with 10,000 access lines, an average community with 20,000 access lines, and a large urban area with 40,000 access lines. Mitchell finds that average incremental cost for residential use is $160 a year for the small urban area, $74 for an average urban area, and $60 for a large urban area (all in 1988 dollars). The average incremental cost of business access and use is about 10 to 20 percent lower. Mitchell's results depend on the average length of loop and busy-hour use. Business subscribers are generally closer to the central office, but they use the network more intensively during busy hours, thus requiring more switching capacity.

The Relationship between Rates and Costs: U.S. Local Exchange Companies

The cost of local access-exchange service depends importantly on the distribution of subscriber loop lengths, the frequency of busy-hour (peak) calling, and technology. For current technology, Mitchell's engineering-economic model appears to provide sufficient detail to give at least a rough estimate of the marginal cost of access and use for the purposes of measuring the departure of current rates from quasi-optimal Ramsey pricing.[13]

10. New England Telephone, Cost of Service Study (COSS), as reported by Palmer, "A Test for Cross-Subsidies in Local Telephone Rates."

11. The cost is a linear function of loop length, obviously an assumption that is too simplistic.

12. Bridger M. Mitchell, *Incremental Costs of Telephone Access and Local Use*, RAND Corporation, July 1990, p. 48.

13. Ramsey pricing is defined more fully below. For the moment, we simply note that Ramsey pricing reflects the most efficient set of prices feasible in a market in which marginal-cost prices do not cover total costs.

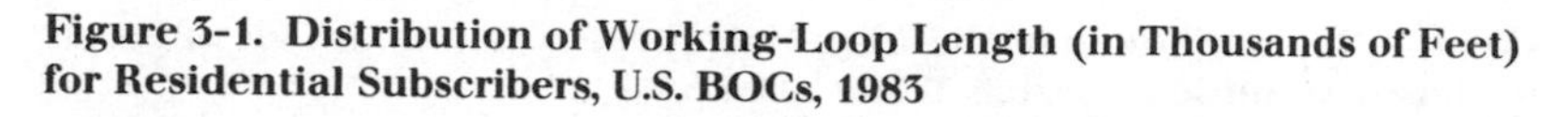

Figure 3-1. Distribution of Working-Loop Length (in Thousands of Feet) for Residential Subscribers, U.S. BOCs, 1983

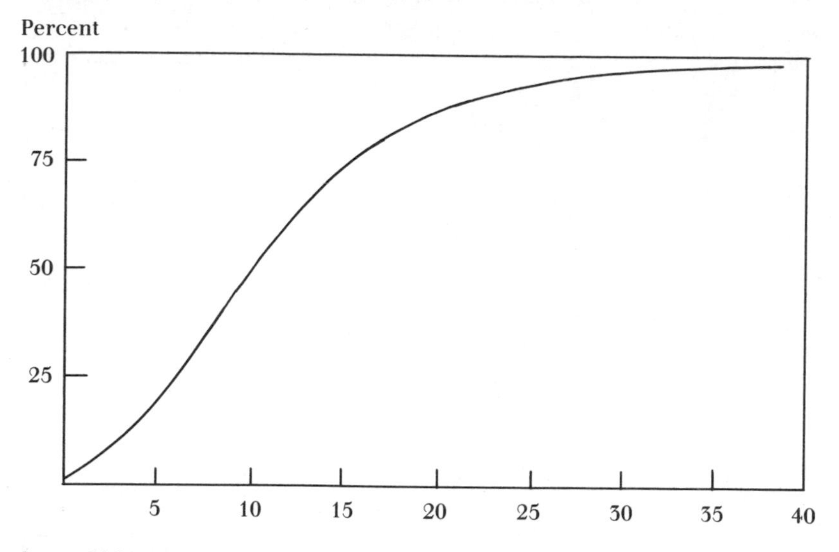

Source: *BOC Notes on the LEC Networks—1990*, issue 1 (March 1991), figure 12-10.

Use of Mitchell's model requires data on the current distribution of U.S. subscribers by length of loop from the central office. These data are available on a summary basis for the BOCs through Bellcore.[14] The most recent data are for 1983. Because the distribution does not change much over time, these data are satisfactory for our purposes. Figures 3-1 and 3-2 show the distribution of working-loop lengths for residential and business customers across the twenty-two BOCs.

The BOC data on loop length may be used to simulate incremental access-use costs for business and residential users in Mitchell's model. This exercise is shown in table 3-1. Average loop lengths for each percentile shown are taken from figures 3-1 and 3-2. Busy-hour use rates and growth-density factors are allowed to vary with loop length in a manner that approximates Mitchell's results for the three model system sizes. The resulting average incremental cost of access and local use evidences substantial increases from the shortest to the longest loop lengths. The incremental costs of residential loops in the ninetieth percentile of

14. Bellcore, *BOC Notes on the LEC Networks—1990*, chapter 12.

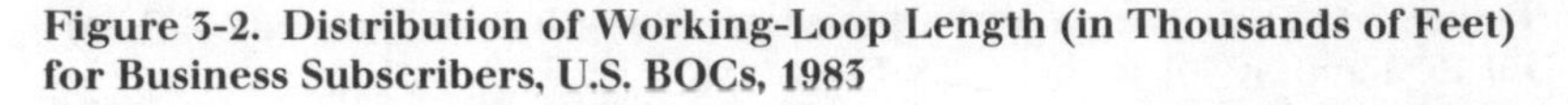

Figure 3-2. Distribution of Working-Loop Length (in Thousands of Feet) for Business Subscribers, U.S. BOCs, 1983

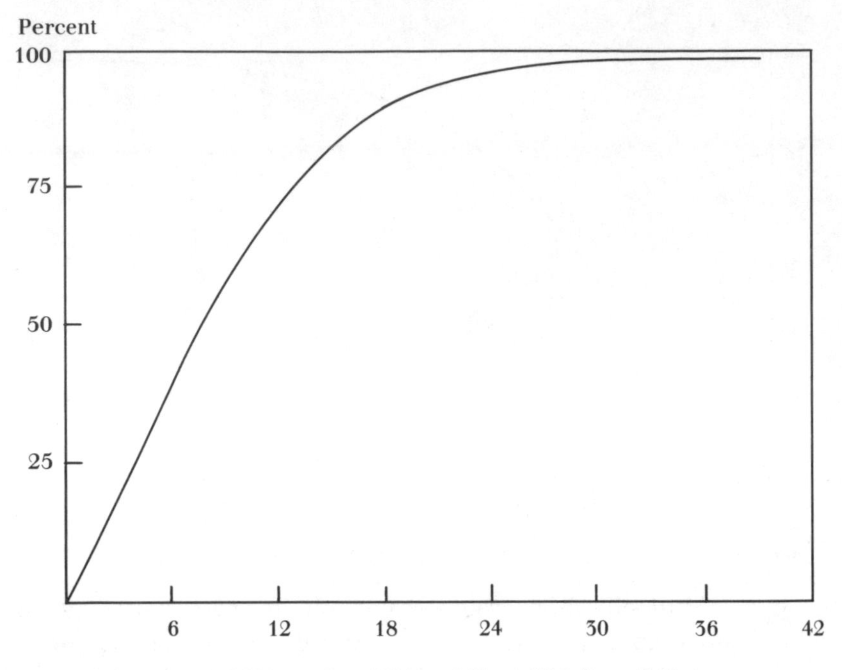

Source: *BOC Notes on the LEC Networks—1990*, issue 1 (March 1991), figure 12-13.

service-area size are $214 a year, or nearly four times the incremental costs of loops in the tenth percentile. Similarly, the incremental costs of business loops in the ninetieth percentile of service-area size are more than two and one-half times the incremental costs of those in the tenth percentile.[15]

To estimate the relationship between loop lengths and local flat access rates for the United States, we assume that the distribution of the number of terminals in a local service area mirrors the distribution of average loop lengths. This is at best a crude approximation, but we cannot improve on it without more detailed data on loop lengths by switching center. Each BOC reports its rates to the National Association of Regulatory Utility Commissioners

15. It should be noted that these estimates are for flat-rate service, including greater use in the larger markets. The differences in the incremental costs of access across service areas would be even greater than those shown in the table because use is assumed to rise with service area size.

Table 3-1. Estimated Average Incremental Cost for Flat-Rate Local Service for Bell Operating Companies, 1990

Percentile	*Loop length (feet)*		*Underground (percent)*		*Calls per month*	*Busy hour*		*Annual cost per loop ($)*
	Feeder	*Distribution*	*Feeder*	*Distribution*		*CCS*	*Attempts*	
				BUSINESS				
90	13,750	3,500	0	0	150	1.8	2.3	161–190
75	9,700	1,000	50	0	175	1.9	2.4	100–127
50	6,800	800	50	100	200	2.0	2.5	82–109
25	2,400	500	50	100	250	2.4	3.0	56–85
10	1,400	400	100	100	300	2.8	3.4	49–81
				RESIDENTIAL				
90	18,000	5,000	0	0	120	1.2	0.9	200–227
75	11,500	3,000	0	0	120	1.6	1.2	137–163
50	8,000	1,500	50	50	120	2.0	1.5	89–115
25	4,000	1,000	50	50	120	2.0	1.5	62–87
10	2,400	600	100	100	120	2.0	1.5	46–71

Source: Based on model developed by Bridger M. Mitchell, *Incremental Costs of Telephone Access and Local Use* (Santa Monica, Calif.: RAND Corporation, July 1990).

Table 3-2. Estimated Average Current Revenue and Average Incremental Cost for Flat-Rate Local Service for Bell Operating Companies, 1990
U.S. $/year

	Business		*Residential*	
Loop-length percentile	*Average incremental cost*	*Average annual revenue*	*Average incremental cost*	*Average annual revenue*
90	175	351	214	152
75	114	386	150	165
50	95	420	102	175
25	70	453	75	184
10	65	462	58	185

Source: Table 3-1 and NARUC (author's calculations).

(NARUC) by size of the local service area. An analysis of these data reveals that in 1990 the average residential rate in the median (fiftieth percentile) service area of the BOCs was $11.06 a month or $133 a year, while the average business rate in the median service area was $29.71 or $357 a year, excluding subscriber line charges. The distribution of rates by service area is shown in table 3-2. Note that the average rate declines as the size of the service area declines and the average loop length increases. This is in sharp contrast to the pattern of estimated incremental access costs, which are also displayed in table 3-2.

The relationship between incremental costs and rates for long-distance access is somewhat more complicated. Mitchell estimates that the annual long-run incremental cost of providing another minute of peak-load capacity for long-distance access (including a tandem switch) is between $7 and $21 a minute. Over a 365-day year, this suggests a busy-hour incremental cost equal to between $0.033 and $0.058 a minute. In addition, each busy-hour attempt requires up to $0.005 in additional costs. Finally, Mitchell estimates billing costs to range from $0.006 to $0.012 per call. The incremental cost of off-peak calls is, therefore, the cost of billing—or no more than $0.012 a minute. For peak hours, however, Mitchell's estimates translate into as much as $0.075 a minute.

In 1990 the Bell operating companies reported $15.2 billion in access revenues and $9.5 billion in switched long-distance revenues. These companies accounted for about 514 billion switched long-distance access minutes in the intrastate and interstate jurisdictions; therefore, they realized nearly five cents a minute of access plus the limited intra-LATA transport required for their own long-distance calls. This is in the middle of Mitchell's range for peak-hour incremental access costs, but it is far above off-peak access costs. It is also far above Perl and Falk's estimate of two cents a minute for toll calls through electronic switches.

Canadian Local-Exchange Services

The distribution of loop lengths for Bell Canada is very similar to that for the U.S. BOCs. Bell Canada's 1981 subscriber loop survey found that the mean working length was 3.102 km.[16] In 1983, Bellcore found that the average working loop length was 3.287 km.[17] Given a similar distribution of loop lengths and a similar technology, it is reasonable to use our estimates of U.S. carriers' incremental costs (shown in table 3-1) for Canada as well.

The rate distortions in Canada are at least as great as those estimated for the U.S. Bell operating companies. A 1992 study by Bell Canada of its Ontario-Quebec service area finds that its long-run incremental costs of access (including use and maintenance) are far above revenues for residential service, particularly in the smaller areas.[18] This is exacerbated by the large extended calling areas used in Canada, which allow free calling over a very large local area. As in the United States, business rates are much higher than residential rates, even though incremental costs are lower.[19] Given the much greater rate differences between rural and urban areas in Canada (see chapter 6), the welfare losses caused by regulation are likely to be much greater per access line than those caused by U.S. regulation.

16. Bell Canada, *Subscriber Loop Survey: Survey Results,* May 1983.

17. Bellcore, *BOC Notes on the LEC Networks—1990.*

18. These estimates were presented in CRTC 92-78 and were derived by Bell Canada from an engineering-accounting framework specified by the CRTC. They showed that the revenues from residential primary exchange service were only 25 percent of attributable costs in the smallest exchanges and only rose to 54 percent of costs in the largest exchanges. By contrast, revenues from business primary exchange service were between 74 and 192 percent of costs.

19. Comparisons between the United States and Canada based on the data in tables 3-2 and 3-3 are difficult because of the differences in the cost concepts. Mitchell's engineering-economic analysis may exclude some of the overhead costs that are included in the Bell Canada PARC study.

Rates and Costs for Long-Distance Services

Most interstate long-distance markets in the United States have been opened to competition.[20] AT&T's share of switched interstate access minutes has fallen to about 60 percent as MCI, Sprint, and other carriers have expanded their national networks. As we have seen, intrastate intra-LATA markets are somewhat less competitive because of the penchant of some state regulators to protect the local-exchange carriers from competition, although such regulatory protectionism is rapidly receding. In Canada competition has only recently been introduced into long-distance markets; therefore, the traditional telephone carriers (the Stentor members) still control about 80 percent of network services.

Given the increasingly competitive structure of long-distance markets, rates may now be declining toward marginal cost of service excluding access charges. Regulated carrier access charges are still likely far above the economic costs of these services, particularly during off-peak hours. In 1993 interstate switched-access charges averaged about 6.75 cents per conversation minute, but they may have been even higher for intrastate markets.[21] In 1993, AT&T paid nearly 43 percent of its total switched long-distance revenues to local-exchange companies for interstate and intrastate access.[22]

Table 3-3 shows the changes in the structure of AT&T's switched interstate rates since 1984, the year of divestiture. In 1984 interstate access charges averaged 17.3 cents per minute. Thus the net revenues from the shortest mileage band were negative (after deducting the access charge) for night and weekend off-peak calls as well as for daytime calls.[23] The taper in the tariff schedule from

20. This does not mean that they are necessarily competitive in the sense that no firm has power over rates and rates approximate long-run marginal cost. See the analysis in chapter 5.

21. FCC, Common Carrier Bureau, Industry Studies Division, *Monitoring Report*, May 1994, table 5.11.

22. FCC, *Statistics of Communications Common Carriers*, 1993–94 edition, pp. 39–41.

23. This is undoubtedly the result of the "Brandon effect." Paul Brandon of Bellcore noted that an access charge system that is based in large part on allocating fixed costs over all long-distance carrier minutes will result in *incremental* access charges

Table 3-3. AT&T Switched Interstate MTS Rates Since Divestiture
(U.S.$ per minute for a five-minute call)

	Gross			
Mileage band	*Daytime 1984*	*Daytime 1993*	*Night 1984*	*Night 1993*
1–10	0.16	0.21	0.07	0.11
4250+	0.55	0.33	0.22	0.17
	Net of access charges			
Mileage band	*Daytime 1984*	*Daytime 1993*	*Night 1984*	*Night 1993*
0–10	−0.013	0.143	−0.103	0.043
4250+	0.377	0.263	0.047	0.103

Source: FCC, *Monitoring Report*, 1984 and 1993.

shortest route to longest route was substantially greater in 1984 than in 1993, both gross and net of access charge.[24] This reduction in the price of distance is a reflection of technological progress and, presumably, of competition, since the cost of transmitting signals over great distances has fallen with the introduction and rapid improvement of fiber optics.

Long-distance pricing has become much more complex with the advent of competition in the United States and, more recently, in Canada. A variety of incentive plans are now offered to residential subscribers that provide discounts based not only on time of day but on cumulative calling. Large business customers negotiate specialized packages that provide a variety of data and voice services, including the connection of their private networks to the switched public network. Nevertheless, there is still a debate about the effect of competition on U.S. interstate rates (see chapter 5).[25] For the present, our major concern is with the effect of *regulation* on the structure of telephone rates, and this effect is registered in interstate long-distance markets principally through carrier access charges in the United States and now in Canada also.

that are inversely related to the carrier's share of the market. In the extreme case of monopoly, this incremental cost approaches zero.

24. It must be recognized that the longer-distance rates have actually declined by more than is suggested in table 3-4 because of various discount plans.

25. This debate has not been resolved because the data on rates, costs, and price-cost margins are far from satisfactory. For a summary of the various studies on this issue, see the appendix to chapter 5, p. 167.

Estimating the Magnitude of Rate Distortions: A Synthesis

We now turn to an attempt to estimate the magnitude of the current static welfare losses from regulatory rate distortions. These losses derive from the failure of local rates to reflect differences in incremental costs and from the general mispricing of local and long-distance service. The static losses are simply the loss in the social value of output caused by underconsumption of long-distance services and overconsumption of local access and use—particularly in high-cost areas. The dynamic costs of mispricing, which we consider later, are those that derive from insufficient investment and competition in bottleneck services. Indeed, the latter costs are generally ignored in the literature, but they could be substantially greater than the static costs of regulatory rate distortions.

Optimal Pricing

The traditional regulatory problem in telecommunications is to set rates so as to maximize the regulators' political-economic welfare function subject to the constraint that embedded costs are covered by the cash flows from all services. It is generally assumed that there are economies of scope or scale that preclude marginal-cost pricing, but there is no consensus on this issue.[26] As the technology of telecommunications changes from one of paired copper wires, microwave towers, and electromechanical switches to a mixture of radio-based systems, fiber optics, local area computer networks, and advanced electronic switching and control systems, most of this historical evidence is likely to become irrelevant.

Although there is a large and growing literature on quasi-optimal pricing of telecommunications services in the presence of scale and scope economies, there is a dearth of evidence on the determinants of actual rates.[27] We begin our empirical investigation by

26. See Waverman, "U.S. Interexchange Competition," for a critique of the empirical evidence on scale/scope economies in communications.

27. For an excellent treatment of this literature, see Bridger M. Mitchell and Ingo Vogelsang, *Telecommunications Pricing: Theory and Practice* (Cambridge University Press, 1991). See also Roger G. Noll and Susan R. Smart, "The Political Economics of

developing a model of the regulator's problem of allocating the fixed, non-traffic-sensitive costs of telephone service across residential and business access, carrier access, and toll rates for local access-exchange carriers.

The carrier's revenues from telephone service derive from monthly connection charges for residences and businesses, long-distance carrier access charges, and toll services:[28]

$$R = P_B Q_B + P_R Q_R + P_A Q_A + P_T Q_T \tag{3-1}$$

where Q_B and Q_R are business lines and residential lines, respectively; Q_A and Q_T are minutes of carrier access—the origination and termination of calls for other long-distance carriers—and toll service, respectively; and the Ps are the regulated rates charged for access or use.

The carriers' costs, C, depend on the number of customer access lines provided and the intensity of the use of its network. Accounting costs are frequently divided into non-traffic-sensitive and traffic-sensitive costs for regulatory purposes. While the actual division is arbitrary, the concept is useful from an analytical perspective. We therefore define costs as:

$$C = C_{TS} + C_{NTS} \tag{3-2}$$

where C_{TS} are traffic-sensitive costs and C_{NTS} are non-traffic-sensitive costs.

Non-traffic-sensitive costs are those that must be incurred to provide subscriber services regardless of traffic volume. They include installation and maintenance of subscriber lines, the connection of these lines to a central office switch, a portion of the switch itself, and the fixed costs of administering and billing for access. These costs depend largely on the number of subscribers, their geographical density—and, therefore, their average distance from a central-office switch—and the technology used in the network:

$$C_{NTS} = C_{NTS}(Q_R, Q_B, M/L, T) \tag{3-3}$$

State Responses to Divestiture and Federal Deregulation of Telecommunications," in Barry G. Cole, ed., *After the Breakup: Assessing the New Post-AT&T Divestiture Era* (Columbia University Press, 1992), pp. 185–234.

28. We exclude directory advertising and billing/collection services from this analysis.

where M is the number of miles of distribution and feeder lines, L is the sum of Q_R and Q_B, and T is a proxy for technology (digital versus analog; fiber versus copper cable).

Traffic-sensitive costs are those that vary with calling volume—billing costs, a portion of switching costs, and the costs of interoffice trunking. These costs vary with the distance of the call and the number of steps in the switching hierarchy required for the call. Except for billing costs, however, these are largely peak-load costs, since the additional switching and trunking costs of calls during off-peak periods are zero. Thus it is difficult to specify any simple relationship between calls and traffic-sensitive costs.[29]

The regulator's problem is to establish a framework that results in rates—the P_is—that maximize its political-economic objective function subject to the constraint that R=C. But what is this objective function?

If the regulator's objective is to maximize economic welfare, it will attempt to set rates that equal incremental costs:

$$P_i = \delta C/\delta Q_i \qquad i = 1,\ldots,4 \tag{3-4}$$

for each of the i = 1, . . . ,4 services, assuming that there are no economies of scale, that services are independent in demand, that there are no consumption externalities, and that there are no economies of scope.[30]

In the presence of scale economies, (3.4) is unachievable if the revenue constraint is to be satisfied. Under these conditions, it is well known that Ramsey prices are required to maximize economic welfare:[31]

$$\frac{P_i - \delta C/\delta Q_i}{P_i} = \frac{-k}{\eta_i}, \tag{3-5}$$

where k is the Ramsey number and η_i is the price elasticity of demand for the ith service.

Assuming that business and residential lines are leased on a flat-

29. Bridger Mitchell concludes that busy-hour use for the average subscriber in his study accounts for between 16 and 23 percent of average incremental costs of a local line. See Mitchell, *Incremental Costs of Telephone Access and Local Use*, p. 48.

30. Note that $\delta C/\delta Q_i$ is equal to $\delta C_{NTS}/\delta Q_i$ plus $\delta C_{TS}/\delta Q_i$.

31. See William J. Baumol and David F. Bradford, "Optimal Departures from Marginal-Cost Pricing," *American Economic Review*, vol. 60 (June 1970), pp. 265–83.

rate basis with at least some minimal amount of free use, the monthly marginal cost of a line should thus include some element of both non-traffic- and traffic-sensitive costs. On the other hand, the marginal cost of carrier access and toll should be composed of only traffic-sensitive costs. The only difference between equations 3-4 and 3-5 is the latter's inverse dependence on the demand elasticities of the various services. Given the results from the empirical literature on telecommunications demand, the departures of rates from marginal costs should be greatest for the business and residential line lease rates and the lowest for toll and carrier access rates.[32] Thus maximization of economic welfare in the presence of scale or scope economies requires that the services with the most inelastic demands bear the highest contribution over marginal cost.

The presence of consumption externalities mitigates somewhat the effects of differential demand elasticities under Ramsey pricing. However, given the near universality of telephone service in the U.S. and Canada, such consumption externalities are rapidly approaching zero.[33] But even if some consumption externalities exist, it is likely that rate reductions from pure Ramsey prices for all subscribers are an inefficient method of accounting for these externalities because an overwhelming proportion of subscribers would not detach from the network in response to a reduction in the number of subscribers or to any conceivable increase in the monthly rate. Thus the Ramsey conditions equation 3-5 may be taken as a rough approximation of optimal prices under 1990s market conditions.

Empirical Estimates of Ramsey Prices

Two components of the repricing of telephone service in the United States and Canada are required to increase economic welfare. First, long-distance rates should be lowered substantially—particularly through the reduction of the contribution charge paid by carriers for access to the local loop. Second, local rates should

32. See Lester D. Taylor, *Telecommunications Demand in Theory and Practice* (Dordrecht: Kluwer Academic Publishers, 1994).

33. See Mitchell and Vogelsang, *Telecommunications Pricing: Theory and Practice,* pp. 55–61, for a discussion of the effect of consumption externalities on Ramsey prices.

be restructured by raising rates in less densely populated areas far more than the rates in urbanized areas. Neither change is likely to be easy in a political environment, although the United States has at least begun the first task.

To estimate Ramsey prices for the United States, we begin with our estimates of the Bell operating company data for incremental access-use costs shown in table 3-1. We assume that long-distance costs for access and short-distance transport are 2 cents per minute.[34] We also assume in the base case that the price elasticity of demand for business access is −0.02, the elasticity for residential access is −0.02, and the price elasticity of demand for long-distance service is −0.75.[35] Finally, we assume that current rates just meet the revenue requirement; thus any new rates must achieve the same revenues. The Ramsey prices from this exercise are shown in table 3-4. The resulting revenue changes are shown for the base case in table 3-5.

Note that in the base case, residential rates in all but the largest cities (with the shortest loops) are required to rise. In some cases the increase is significant—from $12.67 per month to $52.08 per month for the longest loop lengths shown. Business rates decline for all but the longest loops, in some cases significantly. Most important is the fact that access and long-distance rates decline to a very small markup over our assumed two cents per minute incre-

34. We explicitly ignore the welfare gains possible from peak-load pricing. These gains are declining, however, as the cost of peak switching and transmission declines. See Rolla Edward Park and Bridger M. Mitchell, *Optimal Peak-Load Pricing for Local Telephone Calls,* RAND Corporation, March 1987.

35. These elasticities are based on the evidence summarized in Taylor, *Telecommunications Demand in Theory and Practice.* The lowest estimates for the price elasticity of residential demand are those developed by J. Bodnar, P. Dilworyth, and S. Iacono, "Cross-Sectional Analysis of Residential Telephone Subscription in Canada," *Information Economics and Policy,* vol. 3, no. 4 (1988), pp. 359–68. The elasticity of demand for carrier access is based on a study by J. P. Gatto, J. Langin-Hooper, P. B. Robinson, and H. Tryan, "Interstate Switched Access Demand," *Information Economics and Policy,* vol. 3, no. 4 (1988), pp. 333–58. The price elasticity of demand for business lines is probably lower in absolute value than −0.02 in the range of current prices, but we are simulating results that often require sharp local rate increases. Given these changes, we feel that an elasticity of −0.02 is more likely to be accurate, reflecting the greater ability of business customers to find bypass alternatives to the local loop. Throughout, we ignore the possibility that cross elasticities of demand between access and long-distance services are non-zero.

Table 3-4. Ramsey (Quasi-Optimal) Rates for Local U.S. Carriers (BOCs) under Various Price-Elasticity Assumptions, 1990

			Ramsey prices assuming following price elasticities				
			(Base case)	*(2)*	*(3)*	*(4)*	*(5)*
Loop length percentile	*Current ($/yr.)*	Business	−.02	−.01	−.03	−.02	−.01
		Residential	−.02	−.02	−.03	−.02	−.03
		Long distance	−.75	−.75	−.75	−.85	−.50
			BUSINESS				
90	351		510	1,107	512	510	1,267
75	386		332	721	333	332	825
50	420		277	601	278	277	688
25	453		204	443	205	204	507
10	462		189	411	190	189	470
			RESIDENTIAL				
90	152		624	369	625	625	301
75	165		437	259	438	438	211
50	175		297	176	298	298	143
25	184		219	129	219	219	105
10	185		169	100	169	169	82
Long-distance access rate ($/min.)	.048		.0204	.0202	.0205	.0203	.0204
Net static welfare gain from repricing (billion $)			6.42	6.59	6.29	7.68	3.91

Source: See text and table 3-6 (author's calculations).

Table 3-5. Calculation of the Effects of Ramsey Quasi-Optimal Repricing for U.S. Local Exchange Carriers (BOCs), 1990

Item	*Before repricing*	*After repricing (base case)*
Average residential flat rate[a]	173	373
Average business flat rate[a]	305	306
Number of residential access lines[b]	71.9	70.8
Number of business access lines[b]	30.5	30.5
Total local revenues from basic service[c]	21.7	35.8
Average revenue per long-distance minute[d]	0.048	0.0204
Number of access minutes[e]	514	978
Long-distance revenues, including access charges[c]	24.7	19.9

Source: Author's calculations.
a. Dollars per year.
b. Millions.
c. Billion dollars.
d. Dollars.
e. Billions.

mental cost.[36] The welfare gains from this repricing exercise, based on the BOC universe, are $6.4 billion per year. Assuming that the rest of the U.S. local telephone industry has a similar price-cost situation, the potential welfare gains for all of the United States are about $8 billion.

Because estimates of the elasticities are inherently subject to some error, we show alternative simulations based upon higher or lower elasticities. These are shown in the remaining four columns of table 3-4. The most interesting and perhaps obvious result is that a lower price elasticity of demand for business access shifts much more of the repricing burden to business subscribers. The second column of estimated rates shows that a reduction in business demand elasticity to −0.01 results in sharp required rate increases for all but the shortest 25 percent of business loop lengths but more modest increases in residential rates for customers with the longer loop lengths. In addition, the welfare gains depend much more on the assumed elasticity of demand for long-distance and access services than on small changes in the elasticity of demand for business or residential local access. We do not address the optimal design of long-distance tariffs across time periods.

The estimates of the benefits of repricing access and use, shown

36. Obviously, these rates would be higher if actual incremental costs are above our two cent per minute estimate.

in the last row of table 3-4, are derived from a simple model of the demand and cost of providing access and use. The demand functions for access and use are independent; were interdependent demand functions used, the welfare gains would potentially be even larger.

Dynamic Welfare Gains

The static welfare gains from repricing telephone service have been thoroughly studied by others.[37] Missing from these analyses, however, are the potential benefits from competition that could flow if local telephone service were priced efficiently. If residential and business service were priced at Ramsey levels, local service could cost as much as $35 to $55 per month. Residential rates in the smaller communities would be more than trebled. Simply pricing access at long-run incremental cost would require substantial rate increases in the smaller communities. More rational pricing of local access services would surely induce more rapid entry and competition, particularly for those subscribers in dispersed locations. This competition, in turn, could unleash a variety of new services and induce carriers to adopt more efficient technologies. Indeed, as we will argue, it is possible that local telephone service—*properly priced*—is a contestable market in most locations given the preponderance of new technologies to reach subscribers.

Equity

Regulators must be concerned about the equity of large changes in telephone rates. As we have shown, a rational repricing of telephone service could increase residential access rates substantially,

37. The seminal work in analyzing alternative pricing strategies is Park and Mitchell, *Optimal Peak-Load Pricing for Local Telephone Calls,* and *Local Telephone Pricing and Universal Telephone Service,* RAND Corporation, June 1989. Others have estimated the static welfare gains from repricing local and long-distance service, including Jeffrey Rohlfs, "Economically-Efficient Bell-System Pricing," Bell Laboratories, Economic Discussion Paper 138, January 1979; Lewis J. Perl, "Social Welfare and Distributional Consequences of Cost-Based Telephone Pricing," paper presented at the Thirteenth Annual Telecommunications Policy Research Conference, Airlie, VA, April 23, 1985; and John T. Wenders and Bruce L. Egan, "The Implications of Economic Efficiency for U.S. Telecommunications Policy," *Telecommunications Policy,* vol. 10 (March 1986), pp. 33–40.

particularly in smaller communities. These rate increases would cause more than 1 million U.S. residential subscribers to abandon their telephone service, and many of those losing service would be low-income Americans. But many poor subscribers would also *gain* from this repricing because even poor households, particularly those in rural areas, may use a substantial amount of long-distance service.[38] Nevertheless, regulators would want to offset some of the regressive effects of Ramsey-like repricing by offering subsidies directly targeted on lower-income subscribers in the smaller communities. These programs already exist in many states in the form of Lifeline rates. Whatever the cost of such programs to alleviate the burden of repricing on a few million low-income subscribers, it pales in comparison with the estimated $8 billion in annual welfare gains that could be achieved from more efficient pricing of telephone service offered by U.S. local telephone companies.

Conclusion

Regulation is an activity guided by political considerations and justified by policy considerations. Strong consumer and rural lobbies induce regulators to keep access rates low, particularly for rural residences, and use rates high. The policy justification for this practice is the desire to maintain universal service, but it is doubtful whether such rate distortions are necessary to guarantee universal service. The demand for access is extremely inelastic and is likely to be inversely related to the price of use, even long-distance use. Were universality of service with $50 (U.S.) per month access rates in rural areas a serious problem, a Lifeline service for *poor* rural residents would surely be better than the current policy. Moreover, were telephone companies to charge as much as $50 for access through long loops, a wireless service would surely begin to appear as a competitive threat.[39] This development, in turn, might begin to erode the very case for regulating the telephone network.

38. For example, Southwestern Bell has found that residences using its low-income subsidy programs, Lifeline and Link-Up America, spend almost as large a share of their total telephone bill on long-distance services as their average subscriber. See Thomas J. Makarewicz, "The Effectiveness of Low-Income Telephone Assistance Programs: Southwestern Bell's Experience," *Telecommunications Policy* (June 1991), pp. 223–40.

39. See chapter 7.

Chapter Four

New Regulatory Institutions

FROM THE LATE NINETEENTH century until recently, telecommunications carriers have enjoyed explicit or implicit monopoly status. Chapter 2 reviewed recent regulatory actions in Canada and the United States, which have increasingly questioned the economic rationale for the monopoly status of national telecommunications carriers. Nevertheless, since many telecommunications carriers, particularly local-exchange companies, continue to enjoy monopoly status, some process may have to be established to limit the rates charged and the returns earned from the public grant of a monopoly franchise. In the United States and Canada, regulatory authorities oversee the prices charged by telecommunications carriers. Until recently these authorities used rate-of-return, rate-based regulation to control profits and, indirectly, the rates charged by the franchised monopolist. In most other developed countries, telephone service has been offered exclusively by a government-owned firm. In the 1980s, however, privatization and regulation began to replace the state monopolies in many of these countries.

Rate-of-return regulation has come under increasing attack in the past ten years as a costly system inducing neither efficient operation nor socially correct pricing. Partly to remedy these problems, a wide variety of alternative regulatory schemes have begun to appear in the past decade, beginning with the price-caps scheme introduced in the United Kingdom in 1985 to regulate the newly privatized British Telecom.

Rate-of-Return Regulation

We consider a regulatory scheme to be efficacious if it results in a reasonable approximation of competitive conditions with minimum distortions to the firm's operations and if the scheme induces the firm to act in a manner that efficiently meets the social objective of maximizing economic welfare. This chapter focuses on the incentives that the mode of regulation creates for regulated firms to operate in ancillary competitive markets; we see this as a central problem in the current regulatory environment. Does the regulatory scheme induce the regulated firm to diversify uneconomically, to set prices below costs in competitive markets, or to engage in entry-distorting activities? Alternatively, does regulation uneconomically impede entry by regulated firms into new, competitive, or rivalrous markets, thereby depriving the consumers of a source of innovation or additional competitive pressure on prices?

Rate-of-return (ROR) regulation allows the firm to earn no more than its allowable costs, where costs include a predetermined maximum rate of return to be earned by the shareholders who have invested in the equity of the firm. In essence, ROR regulation is a cost-plus contract between the utility and the state. It has a number of defects common to all cost-plus contracts and some that are peculiar to the specific form of the regulatory constraint, namely the control of the rate of return on equity capital.[1]

The disadvantages of ROR regulation that have been well identified in the literature. We do not intend to discuss these items in detail, since they are more than adequately covered elsewhere.[2] Rather, we concentrate on the more important of these handicaps and the possibility that new regulatory approaches might mitigate some of them. Our focus on these problems is in the telecommunications sector, and we ignore many of the issues raised in the

1. Students of rate-of-return regulation have found that the stringency of regulation generally varies with economic conditions. See Paul L. Joskow, "The Determination of the Allowed Rate of Return in a Formal Regulatory Hearing," *Bell Journal of Economics,* vol. 4 (1972), pp. 375–427; Paul L. Joskow, "Inflation and Environmental Concern: Structural Change in the Process of Public Utility Regulation," *Journal of Law and Economics,* vol. 17 (1974), pp. 291–327. See also Paul L. Joskow and Nancy L. Rose, "The Effects of Economic Regulation," in Richard Schmalensee and Robert D. Willig, eds., *Handbook of Industrial Organization,* Volume 2 (Amsterdam: North-Holland, 1989), pp. 1450–1506.

2. See National Telecommunications and Information Administration, U.S. Department of Commerce, *Regulatory Alternatives Report,* July 1987, and John Haring and Evan Kwerel, "Competition Policy in the Post-Equal Access Market," Federal Communications Commission, Office of Plans and Policy, February 11, 1987.

regulation of electricity, natural-gas pipelines, or other less dynamic sectors of the economy.

The first cost of ROR regulation is the cost of the process itself—the costs of record keeping, reporting, and filing; the costs of the regulatory authority; and the costs of the regulatory proceedings themselves. These costs are not small. The British government, in its search for a regulatory process to control the newly privatized British Telecom, considered these pure record-keeping and proceedings costs as "one of the major faults of the ROR system."[3]

More severe problems associated with ROR regulation have to do with the distortions induced by the process rather than the costs of the process itself. The regulatory-economics literature has examined distortions in input-choice, prices, and incentives caused by ROR regulation. ROR affects, among other things, the firm's willingness to accept risk, its choice of debt versus equity, and its allocation of joint costs between regulated and unregulated activities, thereby distorting competition.

INPUT DISTORTIONS. Harvey Averch and Leland L. Johnson demonstrated that a firm regulated with respect to the maximum allowable rate of return on equity had an incentive to be inefficient in both production—to substitute capital for labor (which they labeled rate-base padding)—and pricing. In addition, they showed that firms regulated on a ROR basis had an incentive to misprice services.[4] But these are only part of the harmful effects of ROR regulation.

INCENTIVES. As William J. Baumol has succinctly summarized, ROR regulation involves "substituting the judgments of lawyers for those of business persons and engineers."[5] What becomes important is not customer satisfaction or profits, but playing the regulatory "game," that is, using the regulatory process to the firm's advantage. Again, it is not that the firm is evil or the managers

3. S. C. Littlechild, *Regulation of British Telecommunications Profitability*, Report to the Secretary of State, United Kingdom, 1983.

4. Harvey Averch and Leland L. Johnson, "Behavior of the Firm under Regulatory Constraint, *American Economic Review*, vol. 52, no. 3 (June 1962), pp. 1053–69.

5. William J. Baumol, "Reasonable Rules for Rate Regulation: Plausible Policies for an Imperfect World," in Almarin Phillips and Oliver E. Williamson, eds., *Prices: Issues in Theory, Practice and Public Policy* (University of Pennsylvania Press, 1968), pp. 108–123.

manipulative, but that the process creates incentives for regulatory games and thus the distortion of prices and output.[6] These incentives are particularly evident in telecommunications because technical change increasingly threatens established (regulated) carriers with new entry by firms offering similar or an expanded array of services with new technologies.

An important source of general inefficiency derives from the nature of the cost-plus contract between the state (or customers) and shareholders (or their agent managers), which is at the heart of ROR regulation. The firm is allowed to set prices to recover its costs of operation, including a fair return on the capital invested in the firm. If the firm could charge higher prices in the absence of regulation, it would have little incentive to minimize costs efficiently under perfect ROR regulation. Cost minimization is normally necessary to maximize profits, but it may not be required or even conducive to maximizing profits when a ceiling is placed on the rate of return.

If rate-of-return regulation were instantaneously operational and profit rates measured each second, managers would have no incentives to minimize costs since a dollar saved in costs would not be converted into profits but would instead drive prices down, reducing profits by a dollar. In addition, certain costs may be borne by the firm simply for the benefit of managers—boardrooms, expense accounts, and corporate jets. This is not to argue that the managers of ROR-regulated firms indulge in unlawful acts, but that the incentives to minimize costs are eroded by the inability to keep as profits a dollar of cost reduction that raises the actual return above the allowed return.

In practice, ROR regulation operates with lags and its stringency varies over the business cycle. Therefore, each dollar of cost savings is not instantaneously taken by the regulator, but may be kept by the firm until the next rate review. Ironically, this regulatory lag is essential to keeping ROR regulation from removing incentives altogether.

6. See, for example, Bruce M. Owen and Ronald Brauetigam, *The Regulation Game: Strategic Use of the Administrative Process* (Cambridge, Mass.: Ballinger, 1978); and Roger G. Noll and Bruce M. Owen, eds., *The Political Economy of Deregulation: Interest Groups in the Regulatory Process* (American Enterprise Institute, 1983).

In essence, ROR regulation sets up a direct correspondence between costs and prices: the only means by which a regulated firm can raise its average rates is to raise overall costs. As a result, the regulator has to maintain a bureaucracy that reviews the costs of the firm both to ensure that costs are legitimate and to determine whether costs are current or capital expenses (the latter earning an allowed rate of return). Standardized books of account are set up, and many resources are devoted to duplicating management information and management decisions by the regulator.

It is important to stress that most of the information vital to ROR regulation is under the control of the firm. The firm is capable of manipulating the information to its own advantage. For example, during the initial phases of liberalizing entry into interstate markets in the 1970s, the FCC repeatedly expressed frustration at the inadequacy of the data submitted by AT&T to justify its rate responses to new entry. Regulatory schemes that minimize the information required by the regulator to arrive at a correct decision or that produce incentives for the firm to divulge all the correct information are clearly superior to schemes that give incentives to the firm to hide information. Among the advantages of these newer, incentive-compatible forms of regulation, such as price caps, are the lower level of information required by the regulator and the reduced opportunities for manipulation of data by the firm.

In its analysis of ROR regulation in the U.S. telecommunications sector, the National Telecommunications and Information Administration (NTIA) offered the following judgment:

> Our analysis convinces us that whatever its past virtues and accomplishments, rate base regulation has plainly become an inappropriate mechanism for regulating this rapidly changing industry. It is too costly to implement, requiring large expenditures by regulated firms, public interest and other user groups, and, of course, by regulatory agencies as well. The current process not only entails direct outlays of at least $1 billion yearly—costs that are disproportionate to discernible public interest gains. It also almost certainly imposes even larger indirect cost by discouraging efforts to minimize production costs, dampening regulated firms' incentives rapidly to innovate, and, potentially, facilitating possible anticompetitive behavior.
>
> Ample experience with rate of return regulation demonstrates that, all too often, it proves less an objective process for establishing reasonable prices than a ritualistic game played by firms, regulators, and intervenors—all, ultimately, at the public's expense.

Rate of return regulation, in short, displays pretention of analytical rigor, objectivity, and procedural regularity too often belied by practical realities. For these reasons alone, the public interest would be better served by replacing the current rate of return regulatory scheme by a fairer, more effective, less manipulatable, and less intrusive government system.[7]

THE STRUCTURE OF PRICES. All of the potential distortions of ROR regulation lead to an increased need for regulatory oversight by the very agent whose mode of regulation creates the potential distortions in the first place. Yet the information by which the firm is regulated, as noted, is in the hands of the firm. How are regulators then to judge performance and ascertain whether it is relatively good or bad? The literature that demonstrates that ROR regulation creates distortions suggests that individual rates are often not closely related to costs and are not the prices that are most efficient from the public perspective. As a result, competition in related markets may be impeded or uneconomically stimulated. In some cases, regulated prices may be well above relevant costs (even though prices must be higher than marginal costs if marginal costs are falling), and this pricing umbrella induces inefficient entry. In other cases prices may be below marginal costs, perhaps predatorily, thus inefficiently preventing entry.[8]

Averch and Johnson demonstrated that there may be incentives for a firm regulated on an ROR basis to price below marginal cost in certain markets where demand is highly elastic (say, because of competition). This theoretical finding coincided with the beginnings of competition in the telecommunications industry. In the newly competitive markets, new entrants accused the established firms of predatory pricing practices.[9] As competitors were admitted, regulatory commissions in the United States and Canada began to examine the costs of individual services. This new concern with service-specific pricing and costing led to long proceedings on proper costing methodologies. Regulators struggled with various

7. National Telecommunications and Information Administration, *Regulatory Alternatives Report,* July 1987, pp. 5–6.

8. The results of regulation vary widely across industries. See Joskow and Rose, "The Effects of Economic Regulation," pp. 1464–77.

9. For example, AT&T's Telpak rate for bulk use of private lines, established in 1961, was judged by the FCC to be unjustified (in part) by AT&T's cost studies, based on several cost-allocation methodologies. The issue was never really settled.

approaches for allocating fixed and sunk non-traffic-sensitive costs between different classes of customers. The proceedings became enormously costly and complicated as the regulators introduced and rapidly revised costing rules for individual services.

These protracted attempts at devising allocation rules for apportioning the common fixed costs of service between different classes of customers are unwise and, indeed, futile. Most economists agree that common costs cannot be uniquely and nonarbitrarily allocated among customers and that the average costs that result from such an apportionment procedure, based on historical costs, are likely to result in incorrect prices.[10] Even those participants who see some value in allocation rules consider the present process costly and have been willing to consider alternative rules that provide more leeway for firms while protecting against any potential of predation.[11]

Consider a vertically integrated telecommunications firm, regulated by rate-of-return regulation in some market (say, local service) where a monopoly franchise exists. This regulated firm also operates in other markets, such as long-distance, that are competitive. Averch and Johnson demonstrated that such a firm has an incentive to operate in competitive markets at price levels below attributable costs because the cost of such a strategy may be shifted to the regulated rate payer.

Pricing below attributable costs is predatory behavior under any predation test.[12] Predation in unregulated markets is unlikely since the losses from pricing below cost occur immediately and must be recouped through much higher prices in the future to pay back the investment in predation.[13] Under ROR regulation, however, pre-

10. See Gerald Faulhaber, "Cross-Subsidization: Pricing in Public Enterprises," *American Economic Review*, vol. 65 (December 1975), pp. 966–77; and William J. Baumol, "On the Proper Cost Tests for Natural Monopoly in a Multiproduct Industry," in *Microtheory: Applications and Origins* (MIT Press, 1986), pp. 26–39.

11. Daniel Spullber, *Regulation and Markets* (MIT Press, 1989), pp. 127–31, demonstrates that if attributable costs are correctly estimated for all services, then arbitrary allocations of fixed costs are subsidy free. Hence, these arbitrary allocations do not induce inefficiencies.

12. The definition of predatory behavior is the sale of a product or service below marginal costs in an attempt to drive competitors from the market and subsequently raise prices above marginal costs.

13. In a recent case, *Brooke Group Ltd* v. *Brown and Williamson Tobacco Comp.*, 113 S. Ct. 2578 (1993), the U.S. Supreme Court stated that to demonstrate predation

dation is more likely. A monopoly regulated in market A so as to earn less than its market power would allow can afford to incur losses in competitive markets B and C if these losses can be compensated by price increases in market A. For a regulated monopoly firm, losses in competitive markets may be offset by shifting part of the competitive service's costs to the monopoly market, thereby inducing regulators to allow increases in the rates for the monopoly service.

The regulatory predation story is not complete, however, without asking why a firm would engage in predatory pricing to drive out competitors. Raising rates in the monopoly market has its cost for the regulated firm: the need for a rate filing, the opposition of consumer groups, and the consequent loss of good will with the regulator. There are several possibilities. The regulated monopolist may fear that entry by new firms into adjacent markets could eventually result in threats of entry into its franchised monopoly market, or it may wish to exclude entry into these related markets to exact discriminatory rates across different customers. Alternatively, the regulated firm may be induced to such behavior by the rate distortions erected by regulators because it cannot afford losses in market share in markets in which regulators approve rates far above costs to cross-subsidize other basic or essential services.

Unfortunately, predation cannot be easily detected simply by examining the cost structure of the regulated firm. Investments in new markets may be biased toward those with relatively low marginal costs (or average variable costs) of production that may then, in turn, be used to justify low prices in the new, competitive market.

CONTROLLING PREDATION. The regulatory response to potential or actual predation by franchised monopolists in telecommunications has taken a variety of forms:

---

the plaintiff must prove "that the prices complained of are below an appropriate measure of its rival's costs," and "that the competitor had a reasonable prospect, or ... a dangerous probability, of recouping its investment in below-cost prices." Significantly, the Court noted: "Because the parties in this case agree that the relevant measure of cost is average variable cost, ... we again decline to resolve the conflict ... over the appropriate measure of cost." Elaborating on the second element of this test, the Court said that "for recoupment to occur, below-cost pricing must be capable, as a threshold matter, of producing the intended effects on the firm's rivals," such as "driving them from the market."

—Divestiture (the *AT&T* case);

—Line of business restrictions (in the *AT&T* decree);

—Rules for the structural separation of monopoly from competitive services or accounting-cost separations; and

—Alternative regulatory forms such as price caps.

Rate-of-return regulation has provided other incentives for the monopoly franchise to fight off competition. First, the costs of the regulatory and legal battles themselves are generally permissible expenses in determining the regulatory revenue requirement. It has been the rate payers, not the shareholders, who have been paying to fight competition. Each dollar of expense incurred in reducing competition is generally a dollar of reimbursable cost in a rate hearing. Second, the complex set of cross subsidies that developed under the ROR system created a set of stakeholders who could be armed against the enemy of competition. Thus each attempt to promote competition in telecommunications (such as the *Above 890* decision in 1959 and the *Specialized Carrier* decision in 1971) could be portrayed by the regulated firm as a threat to stable local rates or to the integrity of the network.[14]

Owen and Brauetigam characterize the ROR process as one that gives consumers property rights in prices: prices cannot be altered without an elaborate and expensive hearing.[15] Such a process inevitably evolves into one in which the more politically powerful constituent groups obtain lower rates at the expense of those with less political clout. The alternatives to ROR regulation attempt to avoid this system of cross subsidies and the need for countless hearings.

A number of analyses have been undertaken to estimate the objective function regulators attempt to maximize in rate setting.[16] Economic efficiency is not generally among their objectives. The data seem to support the notion that regulators are motivated to "grease the squeaky wheel." In our empirical examinations of telephone rate setting in Canada and the United States in chapter 6,

14. The evidence of AT&T witnesses in the 1959 hearing was that the licensing of private microwave carriers would cause either "(1) a drastic revision of rate schedules; or (2) great financial harm to the carriers, or (3) a combination of both." 27 FCC at 388 (1959).

15. Owen and Brauetigam, *The Regulation Game.*

16. See Roger G. Noll, "Economic Perspectives on the Politics of Regulation," in Richard Schmalensee and Robert Willig, eds., *Handbook of Industrial Organization,* Volume 2 (Amsterdam: North-Holland, 1989), pp. 1254–87.

we introduce a set of sociopolitical variables to attempt to measure the factors that influence regulators in setting prices. Our attempts, like those of others, are not completely successful; it is difficult to determine the basis on which rates are set under ROR regulation. Nevertheless, regulatory mechanisms that are transparent and do not rely on the unknown objectives of regulators are clearly superior to regulation of the rate base.

EFFICIENT INCENTIVE MECHANISMS. The problem of controlling the monopoly provider of service (or oligopolist with market power) rests on two issues: information and incentives. Regulatory schemes that minimize the regulator's need for information are superior to information-rich schemes. Any regulatory system is a contract between the state (its agent the regulator) and the regulated firm and thus involves a set of incentives. As stressed above, ROR regulation that allows the regulated firm to maximize profits subject to a constraint on the earnings of capital sets up incentives to distort capital choice, underprice certain services, and engage in uneconomic diversification. These incentives of ROR regulation are not compatible with minimization of costs or maximization of social welfare. A regulatory scheme that induces the firm to act efficiently and maximize welfare is said to be *incentive compatible*—the regulation contract has incentives for the firm to choose inputs and outputs efficiently so as to maximize society's welfare. Rate-of-return regulation as depicted here is not incentive compatible.

Efficient Pricing for Public Utilities—The Price-Cap Mechanism

Figure 4-1 depicts a classic natural monopoly with declining unit costs resulting from constant or falling marginal costs and large fixed costs that are common to all customers. (Examples of fixed common costs are the switching offices in telecommunications systems or the local loop to the residential consumer.) The aggregate demand for service is shown in figure 4-1 as DD. At a price (P) equal to marginal cost (MC), the firm loses money. Therefore, prices cannot be set solely on the basis of marginal costs. Regulators thus are forced to allocate common fixed costs and ultimately to base prices on such costs in a rate-of-return process. As stated

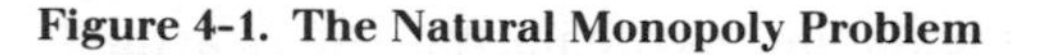
Figure 4-1. The Natural Monopoly Problem

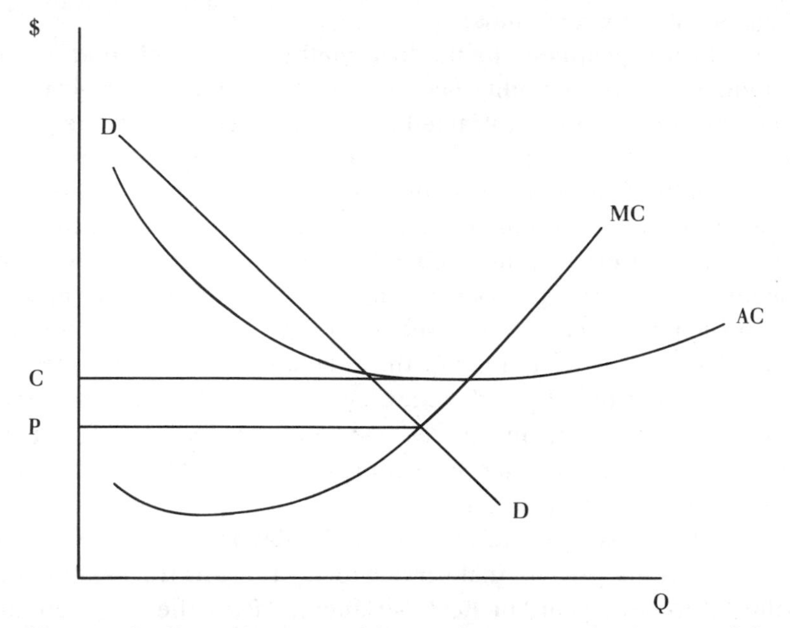

earlier, such allocations are arbitrary, they are not unique, and they do not lead to the prices that optimize resource allocation.[17]

The prices that maximize efficiency are the so-called Ramsey prices. Ramsey prices depend on marginal costs and demand elasticities.[18] ROR regulation does not induce the firm to choose Ramsey prices; thus economists have searched for regulatory mechanisms that lead to both improved input efficiency and a closer approximation of efficient pricing.

To minimize the disincentives for efficient production and to limit the ability of regulated firms to engage in cross-subsidization of services in related, competitive markets, regulators have begun to turn to alternative processes—the most prevalent alternative is called price caps. It is important, however, to understand the lim-

17. Baumol, "On the Proper Cost Tests for Natural Monopoly."

18. Frank Ramsey, "A Contribution to the Theory of Taxation," *Economic Journal*, vol. 37 (1927), pp. 47–61; and William J. Baumol and David F. Bradford, "Optimal Departures from Marginal Cost Pricing," *American Economic Review*, vol. 60 (June 1970), pp. 265–83. Baumol and Bradford provide a more modern treatment. See chapter 3 for the determinants of these Ramsey prices.

itations of this mode of regulation, particularly as it is actually implemented by regulators.

If a firm is regulated by the ROR method, it is likely that it has minimal pricing flexibility between services or customer classes and must engage in a protracted process to raise the average price level. Under ROR regulation, this price level is itself dependent on the costs the firm incurs. Assume that, as an alternative, the firm is allowed to choose individual prices subject to a cap on the average level of prices (and subject to certain constraints discussed below). If the firm lowers costs, it is allowed to keep the profits from such a reduction.[19] The regulator does not set prices in relation to costs, thus eliminating the regulatory link between rates and costs that breeds inefficiency, creates incentives to overcapitalize, and may create incentives to set some prices below marginal cost. This regulation of a price ceiling on regulated prices is called the price-caps (PC) process.

Denote $q_{i,t}$ as the quantity provided today to the i-th customer class, $p_{i,t}$ as the price of that service today, $t + 1$ as the next period (the future test year, in ROR parlance), ΔP as the proportional change in some aggregate (economywide) price index between today and next year (say, the CPI), and X as the productivity effect, the expected annual excess rate of change in productivity for the firm. The price-cap process involves establishing the following constraint on the firm's rates:

$$\text{(4-1)} \qquad \frac{\Sigma\, p_{i,t+1}\, q_{i,t}}{\Sigma\, p_{i,t}\, q_{i,t}} \leq 1 + (\Delta P - X).$$

The PC rule is thus a Laspeyres price index that may increase by no more than $\Delta P - X$. The firm may change individual prices, but the average price is not allowed to increase by more than the rate of general inflation less the cost-reducing effects of firm-specific productivity improvements, both static and dynamic.

PRICE CAPS AND INPUT DISTORTIONS. Under the pure form of PC regulation depicted here the firm has no incentive to substitute artificially one input (such as capital) for another (such as labor). The reasoning is simple. Under ROR regulation, inefficiently sub-

19. This ability may be reduced by any subsequent review of whether the cap is inadequate or excessive. See below.

stituting capital for labor increases the rate base, thus increasing the total revenue requirement. Under PC regulation, inefficiently substituting capital for labor does not increase allowable revenues (which are based on prices, not costs), but instead only decreases profits. As a result, such a substitution is not profitable. Input distortions are not induced by PC regulation.

PC regulation thus provides the correct incentives for cost minimization. A dollar of increased unnecessary costs cannot be collected from customers (as they can in ROR regulation); increased (and unnecessary) costs only lower profits under a PC rule. Unnecessary costs thus will be avoided by a firm under PC regulation. An important corollary follows—there is little need for the regulator to examine judiciously the firm's costs, and thus the informational requirement is lessened. Under ROR regulation the regulator is forced into such examinations.

PRICE CAPS AND CORRECT PRICES. Under the PC process the average real price of the service must fall each year by the formula given in equation 4-1—the average price rises by the rate of inflation less the productivity offset. Vogelsang and Finsinger have shown that such a constraint on the firm induces the firm to choose quasi-efficient Ramsey prices over the long term.[20]

A simple example of the efficiency-pricing effect of price caps may be useful. Two services are offered under ROR regulation, each at ten units of demand. The price of the first service is one dollar below costs; the other price is one dollar above costs. The firm breaks even. Assume that the service priced below cost is subject to price-inelastic demand, and assume that the demand for the service priced above cost is highly elastic. Raising the price of the service priced below cost will raise profits; lowering the price of the service priced above cost will also raise profits since the quantity demanded increases proportionately more than the price is lowered.

A profit-maximizing ROR-constrained firm has no incentive to make these price changes since it would then violate the profit ceiling. A PC regulated firm is not so inhibited. As long as the price changes offset each other so as to satisfy the regulatory price con-

20. Ingo Vogelsang and Jorg Finsinger, "A Regulatory Adjustment Process for Optimal Pricing by Multiproduct Monopoly Firms," *Bell Journal of Economics*, vol. 10 (Spring 1979), pp. 157–71

straint, increases in profits are not subject to regulatory confiscation. The price-capped firm will raise prices where demand is relatively inelastic and lower prices where demand is relatively elastic. If the regulatory PC constraint reduces profits, in the next period the firm continues to adjust prices in this process. But these price adjustments are simply steps along a price path determined by equation 4-1. Thus PC regulation ultimately leads the firm to set socially correct prices (Ramsey prices) for individual services.

Making Price Caps Operational

A number of jurisdictions now use a PC-type process: Britain; the FCC for AT&T[21] and the local-exchange carriers; New Zealand; and several U.S. states. The FCC first proposed to replace ROR regulation of AT&T with PC regulation in 1987, finally adopting price caps in March 1989. Under this process, AT&T's prices were divided into three baskets of services—residential and small-business services, 800-number services, and other business services (the competitive offerings of AT&T).[22] Each basket was treated separately so that a price increase in a monopoly service could not be used to cross-subsidize a competitive service offering.[23] In addition, limits on maximum increases and decreases for a number of individual services were incorporated. X—the productivity offset—was set at 3 percent per year. The scheme was set to operate from 1989 through 1993, with an evaluation of its effects to be conducted in 1993. It incorporates a Laspeyres price index, so that for any basket:

$$\text{(4-2)} \qquad \frac{p_{t+1} \cdot q_t}{q_t} \leq p_t \,(1 + \Delta P - X).$$

If the proposed set of prices in equation 4-2 meets the cap, a streamlined approval process is set in place.[24]

21. Now that AT&T has been declared a nondominant carrier in the U.S. interstate market, the FCC's price-cap regime for AT&T has been all but abandoned.

22. In 1991 and 1993, baskets 2 and 3 were held to be competitive, leading to their removal from price-cap regulation.

23. This approach also limits the ability of the regulatory carrier to move rates toward Ramsey prices.

24. For a discussion of this approach, see David E. M. Sappington and David S. Sibley, "Regulating without Cost Information: The Incremental Surplus Subsidy Scheme," *International Economic Review*, vol. 29 (May 1988), pp. 297–306.

While simple PC regulation theoretically has efficiency properties that are superior to simple ROR regulation, the transition from ROR to PC regulation overturns some of the advantages described above. The purpose of the PC process is to divorce the regulated firm's average revenue decision from its average cost decision. The incentive mechanism is profits. However, in setting X, the productivity offset, regulators inevitably must examine profits. Therefore, knowing that regulators will subsequently use its actual profits to set X in the future, the firm can act strategically by being inefficient or choosing socially incorrect outputs. Thus some authors have suggested a correspondence between price caps and ROR regulation with a four- or five-year regulatory lag.[25]

A substantial gap can exist between the theoretically simple pure PC model discussed above and its actual implementation. Several recent papers extend the PC framework to deal with realistic issues. D. E. M. Sappington and D. S. Sibley examine the actual FCC PC average revenue formula for a firm that uses two-part tariffs.[26] They show that the firm so constrained will engage in strategic pricing, raising the fixed part of the two-part tariff (the access fee) and lowering the use fee. Lowering the use fee increases demand, allowing the fixed (access) fee to be raised in the future. Using linear demand and costs and a four-year process (the actual FCC PC process), Sappington and Sibley find that it is likely that the actual PC process lowers consumers' surplus while producers' surplus increases, as does total welfare.

Two alternative formulations used in the Sappington-Sibley analysis alter the distribution of the gains in moving from ROR to PC regulation. One relies on a constant previous-year weighted index, where the weights are not last period's quantities but some base year (say, q_o) that is used in all subsequent periods. This base-year constant weighting essentially limits the increase in the access fee. A second formulation keeps the original tariff as an option in subsequent periods. In this formulation, consumers' surplus must increase (or else no consumer would choose the new prices).

25. John Vickers and George Yarrow, *Privatization: An Economic Analysis* (MIT Press, 1988). The authors include asymmetric information and the rate of return on capital as the regulatory target under both schemes.

26. David E. M. Sappington and David S. Sibley, "Strategic Nonlinear Pricing under Price-Cap Regulation," *RAND Journal of Economics*, vol. 23 (Spring 1992), pp. 1–19.

Pint examines both ROR and PC models in a model of stochastic cost.[27] Both processes lead to an underinvestment in effort (managers do not maximize efficiency) and overcapitalization. In the Pint model, a PC process leads to overcapitalization because the firm knows that its performance is to be examined in a future year. The longer the time between reviews, the lower the firm's incentives to distort. A four-year regulatory cycle that uses average costs for the entire four years has better results (superior social welfare and higher consumers' surplus) than a review that uses only the last year's data (or some test year's data), as in the FCC process.

The major practical issue in PC regulation is the adjustment of the X factor over the years in a political environment in which the regulators are criticized for allowing the regulated firm to overearn. Moreover, such a result may result in a literal violation of the regulators' statutory mandate to ensure that rates are just and reasonable. As a result, regulators are rarely able to implement a pure PC process, preferring instead to attach a banded rate-of-return provision that requires the carriers to adjust rates when their earned rates of return exceed or fall below the limits of the band.[28]

The Price-Caps Process and Competition

Does the price-caps regulatory rule represented in equation 4-2 prevent predation and uneconomic diversification? Assume that all services, monopoly and competitive, are aggregated in one basket. Consider a basket of two services, 1 and 2. The first is a monopoly franchise service for which the existing price constrains the firm—the present price, p_1, is below the monopoly level, p_1. Price p_2 is set at the competitive level, and the firm meets the regulatory constraint in period 1:

$$(4\text{-}3) \qquad \frac{p_1^2 q_1^1 + p_2^2 q_2^1}{p_1^1 q_1^1 + p_2^1 q_2^1} \leq 1 + \Delta P - X.^{29}$$

To simplify the analysis, assume that the productivity offset, X, is

27. Ellen M. Pint, "Price-Cap versus Rate-of-Return Regulation in a Stochastic-Cost Model," *RAND Journal of Economics*, vol. 23 (Winter 1992), pp. 564–78.

28. The alternative is to adjust the X factor in response to earnings realizations, as the FCC has recently done for the interstate (access) rates of the regional BOCs. Either approach violates the goal of separating cost realizations from approved rates.

29. The superscripts 1 and 2 represent the first and second periods, respectively.

just equal to the inflation rate, ΔP, and that the regulatory constraint is satisfied as an equality (that the firm just meets the constraint). Then, if the regulated firm attempts to engage in predation, it must satisfy:

$$(\Delta p_1)q_1^1 = (\Delta p_2)q_2^1 \qquad \text{(4-4)}$$

Now refer to figure 4-2. If the firm lowers the price of service 2 to a level below marginal cost, MC_2, it will lose profits equal to the shaded areas, shown as 4 + 5. Its increase in the price of service 1 will generate an increase in profits of area 1 less area 2.[30] But the price-caps constraint requires that areas 1 + 3 equal area 4. For current-period profits to be positive from such predation, area 1 − area 2 would have to be greater than areas 4 + 5. Substituting the regulatory constraint, 4 = 1 + 3, leads to the impossible requirement for predation: $-2 > 3 + 5$. Thus predation cannot be a successful strategy under price caps in the short run.

Price caps can, however, reduce the cost of predation. In the absence of regulation, a firm producing both services would be maximizing its profits from service 1 regardless of its activities in the market for service 2.[31] It would, therefore, be forced to bear the full brunt of the losses from predation, areas 4 and 5 in figure 4-2. With rate caps, this net loss is reduced by the area 1 − the area 2 because the firm can raise the price of service 1 that is restrained by regulation. Thus predation is more likely under rate caps than in an unregulated private market, but it is less likely than under standard ROR regulation. For this reason, it is important to segregate competitive and monopoly services under a price-caps regime, keeping competitive services out of the regulatory regime altogether.[32]

Therefore, the choice of baskets and the division of services into competitive and monopoly areas is critical. It is unlikely that devising baskets is as treacherous an area as the issue of cost separation and structural separations in the ROR arena for several reasons.

30. This may or may not be positive, of course, depending on the price elasticity of demand in the area around p_1^1, q_1^1.

31. Assuming that the two services are independent in production and in demand.

32. Regulators are likely to want to segregate services into baskets for quite a different reason: to prevent profit-maximizing regulated firms from moving rates toward Ramsey (quasi-optimal) levels, generally in the name of universal service.

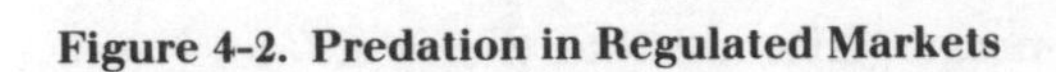

Figure 4-2. Predation in Regulated Markets

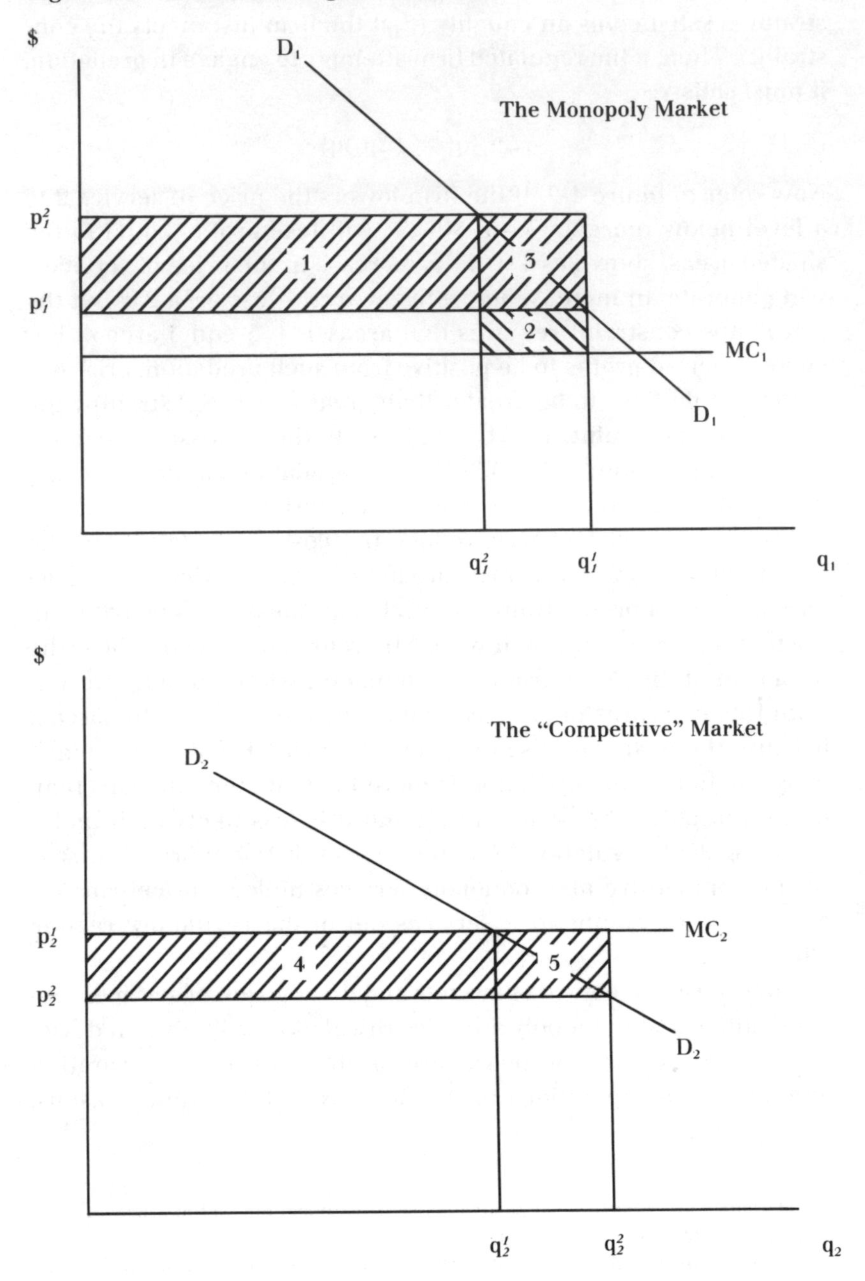

First, the number of products subject to competition is substantially smaller than the set of all products offered by the firm. Second, deciding whether a service is competitive is far simpler than devising cost separations for all products. Third, rearranging baskets is not complex.

Other Regulatory Schemes

A host of regulatory schemes that remove some of the regulatory burden while allowing increased price flexibility have been introduced. The theoretical literature has not examined the efficiency of most of these processes. These regulatory schemes can be divided into four categories: banded pricing and banded ROR; downward pricing flexibility; earnings sharing and flexibility; and social contracts. We discuss the effects of these schemes on pricing in competitive markets.

Banded pricing schemes are those that allow the firm's rate of return to deviate from the prescribed rate of return before rates must be adjusted through a regulatory hearing or some other device. Circumstances can change, creating either positive or negative surplus in the firm's earnings without triggering a regulatory examination. In addition, the firm can gain the benefits of cost efficiencies that allow the earned rate of return to hit the upper band. The pricing of individual services is, however, still regulated in banded ROR systems. Therefore this scheme, unlike price caps, does not sanction rate rebalancing. Incentives for predation that would exist under a pure ROR system are reduced somewhat under a banded-returns system since a loss lowering the actual ROR will lead to a hearing only if the lower band is breached.

Banded pricing systems permit some pricing flexibility. In theory, banded pricing is not equivalent to a price-caps process since one goal of a price-caps process is the readjustment of prices to their Ramsey levels. In practice, the price-caps processes actually implemented include limits on price increases and price reductions, making these schemes similar to banded pricing schemes.

Downward pricing flexibility without allowance for corresponding price increases elsewhere would appear to minimize predatory behavior since the losses from price decreases would not be compensated by price increases elsewhere. However, the overall ROR is still a significant issue, and decreases in the overall return can still trigger general rate reviews.

An example of earnings sharing and flexibility is found in California. The California Public Utilities Commission (PUC) launched a variant of earnings sharing in 1993 after it had determined that the cost of capital in 1990 was 11.5 percent. Pacific Bell and GTE may now earn up to 13 percent or down to 9.75 percent (for two years in a row) with no general review of rates. Any increase in the rate of return between 13 percent and 16.5 percent will be shared 50:50 with consumers through a credit on basic service. Monopoly services are indexed each year in a process similar to price capping, and prices for competitive services are to be deregulated. Cost accounting separations based on fully allocated costs are in the process of being introduced. The California PUC monitors financial performance, investment, modernization, and quality of service.

The California PUC considers this process more efficient and expects it to induce the cost efficiencies that would be generated by a price-caps process. However, the price-caps process has two purposes: one is a superior form of regulatory oversight; the other is an inducement for welfare maximizing prices. The degree of regulatory oversight in the California process does not appear consistent with the theoretical PC process and thus may lose some of the benefits of price caps. Predation is discouraged by allowing the rate of return to move downward without triggering a rate increase elsewhere. Hence losses in some markets cannot be compensated by gains elsewhere.

The social contract process involves the firm's trading off flexibility of pricing in some sectors; for example, gaining flexibility in long-distance pricing in exchange for rate freezes in other areas such as local service. In states such as Vermont, regulated carriers have entered into such agreements to maintain basic rates in return for partial or full deregulation of other services, such as central-office, business, and information.

Cost Separations and Structural Separation

In some instances, regulators have grappled with the issue of the array of monopoly and competitive services offered by regulated telecommunications firms by devising Chinese walls to separate competitive from monopoly services.

The issue of separating costs in telecommunications regulation has its origins in AT&T's early responses to the *Above 890* decision in the early 1960s. In response to private-microwave competition, AT&T filed a set of tariffs, called "Telpak," that were challenged as noncompensatory. The FCC was forced to determine whether these rates covered AT&T's costs, and this process led to decades of attempting to allocate AT&T's network costs across its various service categories. Were variable costs X attributable to service i or j? How were fixed costs to be allocated (since it was felt that tariffs must return a contribution to fixed costs for the firm to break even)?

The hearings simply demonstrated that any costing rules developed "are of necessity arbitrary in nature and lead to costly and lengthy debates over the fairness of allocations of joint and common costs among the monopoly and competitive operations of regulated firms. . . ."[33] These allocations require constant audit, straining the resources of regulator, regulated, and intervenors.

Structural separation requires the regulated monopolist to establish a separate subsidiary to operate competitive services. This is cost separation taken to the limit—the firm separates those costs (inputs) that are used for competitive services into a separate firm. Two large issues arise in such separations. First, are all inputs (such as, for example, management) to be separated? Second, structural separation eliminates the potential for economies of scope (operating competitive and monopoly services within one firm) and economies of scale (for example, if both competitive and monopoly services use separate inputs that, if combined, would yield scale economies). Is the gain in welfare in competitive markets worth the sacrifice of these economies?

Price-caps regulation can solve the predation problem through service separation—separating services into monopoly and competitive with separate baskets for each. In our example above if q_1 and q_2 were put into separate baskets with separate caps, the firm could not raise p_1 without limit to compensate for decreases in p_2.

Thus a price-caps process with a judicious choice of service baskets can eliminate most incentives of predation. Many regula-

33. Canadian Director of Investigation and Research, Written Argument in CRTC 92-78, Review of Regulatory Framework, January 17, 1994, p. 22.

tors also insist on price floors even within a competitive service basket for firms with market power. For example, the AT&T price-cap order set average variable cost, including all access charges and attributable billing and collection charges, as the floor.

The Prospects for Further Regulatory Reform

While these newer regulatory schemes are superior to rate-of-return regulation, the amount of regulatory oversight and litigation does not appear to decline as much as one would have expected when ROR regulation is replaced with a price-caps process. Regulation consists of much more than choosing the correct price. Regulation becomes a process where enormous gains or losses can be earned by the winners and losers in the process. The regulatory market pits the success or failure of firms not against competitive market standards and the thousands of day-to-day decisions involved—products, prices, marketing—but in a battle for a regulatory decision.

Adam Smith suggested that businessmen rarely meet but to fix prices.[34] A modern Adam Smith might suggest that businessmen prefer the regulatory market over true competition since the regulator has the power to affect the market outcome without exposing the firm to the risk of antitrust prosecution. Even with the perfect price-caps process, we would still view regulation as pernicious. Recent events amplify our view.

First, AT&T has successfully appealed to the courts to overturn the FCC's decision to forbear from requiring MCI, Sprint, and other nondominant carriers to file their tariffs. Thus all long-distance carriers must now file tariffs, creating the possibility of more coordinated pricing among the three large carriers.[35] Second, the need to have true competition in long-distance and other services requires that all firms, including local carriers, offer equal access, open network architecture (ONA), and collocation because net-

34. Adam Smith, *An Inquiry into the Nature and Causes of the Wealth of Nations*, 3rd ed. (London: Methuen, 1922).

35. The allegation of "tacit collusion" among these three carriers has been raised by Paul W. MacAvoy, "Tacit Collusion under Regulation in the Pricing of Interstate Long Distance Telephone Services," *Journal of Economics and Management Strategy*, vol. 4 (Summer 1995), pp. 147–85.

work interconnection is a prime requisite of telecommunications. Competition cannot flourish unless competitors can connect with other networks because the final subscriber does not wish to have separate connections with each and every network through separate devices located in his home or establishment. The difficulties in implementing rules ensuring true competition in the regulatory arena signal how regulatory processes cannot substitute for competition (see chapter 6 for a fuller discussion). Third, even under price-caps rules in some states, effective competition does not exist in intra-LATA toll, because of a variety of other impediments to competition erected by state regulators. We now turn to an empirical analysis of these issues.

Chapter Five

Competition in the Long-Distance Market

UNTIL 1984, telecommunications services in the United States were generally provided by vertically integrated companies. AT&T provided all local services in its franchise areas and more than 90 percent of switched message-telephone (MTS) and private-line services, both interstate and intrastate. AT&T was broken up in 1984 because the Justice Department argued that competition in the long-distance market had been impeded by vertical integration. Mechanisms such as cost separations or regulatory structural separations were considered insufficient at that time to prevent vertically integrated AT&T from using its ownership of the "bottleneck" local facilities to thwart competition in long-distance service and manufacturing.[1] (In Canada, by contrast, the vertically integrated telephone companies continue to provide virtually all local services and almost 80 percent of long-distance telecommunications).

The U.S. Long-Distance Market

Competition in U.S. long-distance markets appears to have escalated over the past two decades. Since the 1984 divestiture, AT&T has steadily lost market share in interstate switched services as other carriers have obtained equal access to local telephone circuits. In 1984, AT&T had more than 80 percent of interstate

1. As Judge Greene opined at the close of the government's case: "The Court finds that as of now, sufficient evidence has been adduced to dictate the conclusion that AT&T has monopolized the intercity services market by frustrating the efforts of other companies to compete with it on a fair and reasonable basis." *U.S.* v. *American Tel. & Tel. Co.*, 524 F. Supp. 1336 (D.D.C. 1981).

switched minutes; by 1994 its market share had fallen to less than 60 percent.[2] In intrastate markets, competition is less intense because state regulatory commissions have been more reluctant to open long-distance service—particularly intra-LATA service (within local access and transport areas)—to new entrants. Recently, however, competition has been growing even in intra-LATA markets.[3]

The introduction of competition in long-distance services focused attention on the pricing of carrier access. Before the 1970s, an integrated monopolist—AT&T—simply shifted revenues among jurisdictions, with the blessing of regulators but out of public view. Industry participants may have known that these "separations and settlements" practices shifted local-access costs onto long-distance use, but few others understood the practice. Once long-distance carriers and local exchange-access carriers were fully separated, however, the revenue shift could be accomplished only with formal access charges paid by the long-distance carrier to local telephone companies for originating or terminating their calls. This brought the debate over the proper level of long-distance contribution to the fixed costs of the local-exchange plant into full view. The ensuing controversy has still not been resolved.

Federal Regulation of Interstate Services

The Federal Communications Commission (FCC) launched a controversial process of repricing telephone service through its decision to impose monthly subscriber line charges (SLCs) on end users to defray part of the non-traffic-sensitive (NTS) costs of local connections assigned to interstate long-distance service. But because of political pressure and the FCC's lack of jurisdiction over intrastate rates, this move toward repricing is far from complete. More important, the FCC can do little about the other regulatory rate distortions in the pricing of local access, including discrimination against business subscribers and subsidies to rural subscribers. These, too, are currently the domain of state authorities.[4]

2. Federal Communications Commission, *Statistics of Communications Common Carriers*, 1993–94 edition, p. 311.

3. See chapter 6.

4. The possible preemption of state authority in some of these matters has been the subject of recent proposed federal legislation.

Table 5-1. Interstate Charges by Local Telephone Companies to Long-Distance Carriers, 1984–94

National average for "premium" service, cents per minute

	Carrier common-line charge			
Period	*Originating access minute*[a]	*Terminating access minute*[a]	*Total traffic-sensitive charge per access minute*[b]	*Total charges per conversation minute*[c]
05/26/84–01/14/85	5.24	5.24	3.1	17.3
01/15/85–05/31/85	5.43	5.43	3.1	17.7
06/01/85–09/30/85	4.71	4.71	3.1	16.2
10/01/85–05/31/86	4.33	4.33	3.1	15.4
06/01/86–12/31/86	3.04	4.33	3.1	14.0
01/01/87–06/30/87	1.55	4.33	3.1	12.4
07/01/87–12/31/87	0.69	4.33	3.1	11.5
01/01/88–11/30/88	0.00	4.14	3.1	10.6
12/01/88–02/14/89	0.00	3.39	3.0	9.6
02/15/89–03/31/89	0.00	3.25	3.0	9.5
04/01/89–12/31/89	1.00	1.83	3.0	9.1
01/01/90–06/30/90	1.00	1.53	2.5	7.8
07/01/90–12/31/90	1.00	1.23	2.5	7.5
01/01/91–06/30/91	1.00	1.14	2.4	7.2
07/01/91–06/30/92	0.88	1.06	2.4	7.0
07/01/92–06/30/93	0.79	0.95	2.4	6.8
07/01/93–06/30/94	0.88	1.16	2.2	6.7

Source: Federal Communications Commission, *Monitoring Report*, May 1994, p. 386, table 5-11.

a. Before April 1, 1989, the rates shown are nationally uniform "premium" rates specified in tariffs filed by the National Exchange Carrier Association (NECA). Since then, the terminating rate is no longer identical for all companies, and the rate shown is an average. Where equal access is not available, carriers other than AT&T pay discounted "nonpremium" rates.

b. Traffic-sensitive switched-access rates are not subject to mandatory pooling and are thus not nationally uniform. The rate shown in this column has been estimated by FCC staff as a weighted average that includes both switching and transport charges.

c. Sum of originating, terminating, and traffic-sensitive charges, including allowance for uncompleted calls.

Since mid-1991 the federal SLC for residential and single-line businesses has been capped at $3.50 per month, or about two-thirds of the interstate portion of the average NTS cost of a local loop. For multiline businesses, the average SLC has been capped at $6.00 per line but cannot go above the average NTS cost per line for the entire state jurisdiction. This maximum limitation resulted in an average SLC for multiline businesses of $5.45 per month in late 1994.[5] The FCC has required that all SLC revenues be used to reduce the interstate access charge for interstate carriers. In 1984, this rate averaged about 17 cents per minute for "premium" interstate access; by mid-1994, it was only 6.7 cents per minute (table 5-1). However, political pressures have kept the FCC from continuing to shift the burden of NTS costs from interstate toll to the final subscriber.

The shift from traditional rate-based regulation of AT&T's interstate operations to price caps in 1989 may also have added downward pressure on interstate rates. Until 1995, AT&T was regulated as a dominant carrier, partly out of fear that it would lower rates so far that many of the new firms might be put in jeopardy.[6] The substitution of price caps for rate-of-return regulation reduced the possibility that AT&T might engage in predatory activities subsidized by regulated services. (See our discussion in chapter 4.)

State Regulation of Intrastate Services

Regulatory policies differ substantially across the three different U.S. long-distance or "toll" markets.[7] The first of these, the interstate market, is regulated by the FCC. The AT&T antitrust decree has mandated equal access in this market, and the FCC has, until recently, regulated the rates of the dominant carrier, AT&T, through price caps.[8] The regional Bell operating companies

5. Federal Communications Commission, Common Carrier Bureau Industry Analysis Division, *Monitoring Report,* May 1995, table 5.10.

6. A report by Peter Huber and associates argues that the new other common carriers (OCCs) have much higher costs than AT&T and therefore cannot compete with AT&T without explicit FCC protection. See Peter W. Huber, Michael K. Kellogg, and John Thorne, *The Geodesic Network II: 1993 Report on Competition in the Telephone Industry* (Washington: Geodesic Co., 1992), chapter 3.

7. We do not use the term *market* in its strict antitrust sense here. We are referring to the separation induced by regulation.

8. Recently through court intervention, the FCC has been forced to require non-

(RBOCs) are not allowed into this market because of the line-of-business restrictions in the modified final judgment (MFJ). The second, the intra-LATA market, is controlled by state regulators. ("Intra-LATA" refers to toll calls wholly within a LATA.) As we have shown, regulation in the intra-LATA market varies widely across the United States, ranging from virtual deregulation to total restrictions on non–telephone company entry and no pricing flexibility.[9] The RBOCs and other local-exchange carriers (LECs) are major suppliers in this market. The third market is the intrastate inter-LATA market. The states regulate these markets—specifying, for example, the degree of price flexibility allowed or the need for rate-of-return hearings. The service providers in the inter-LATA market are AT&T, MCI, Sprint, and other national or regional carriers, all of which now operate under equivalent conditions of access. The RBOCs are also banned by the MFJ from providing service in the intrastate inter-LATA market.

Unfortunately, the states have been less concerned about eliminating the burden of NTS costs on intrastate toll calls. Since 1983 average intrastate MTS rates have declined at a rate of only 0.8 percent per year, while interstate MTS rates have fallen by 2.4 percent per year in nominal dollars (table 5-2).[10] This is not to say that all states have been resistant to repricing intrastate tolls, but most have been far less aggressive than the FCC. For instance, although all multi-LATA states now allow competition in inter-LATA service, four states still do not allow intra-LATA competition, and none had mandated equal access for all competitive carriers before 1995. Moreover, intra-LATA rates are still tightly regulated in many states.[11]

The pace of change in AT&T's interstate and intrastate inter-LATA rates is shown in table 5-3. Interstate rates generally have fallen more rapidly than intrastate rates. The 1983 intrastate rates had a much greater taper from the shortest mileage band to the

---

dominant carriers to also file tariffs. In 1995, however, the FCC ruled that AT&T is "nondominant" in domestic service markets.

9. See chapter 2.

10. We should note that the consumer price index for interstate long-distance service generally understates the rate of decline in rates because it does not adequately cover competitive carriers and fails to account for the myriad of discount plans now in use. For these reasons, the Bureau of Labor Statistics (BLS) is currently undertaking a major revision in its telephone service price indexes.

11. See chapter 2.

Table 5-2. U.S. Producer Price Indexes for Intrastate and Interstate Long-Distance Message Telephone Service, 1977–94
(1972 = 100)

Year	*Interstate*	*Intrastate*
1977	120.6	131.9
1978	121.9	132.0
1979	120.8	131.6
1980	124.6	132.3
1981	137.5	137.3
1982	152.0	145.6
1983	153.4	152.1
1984	148.8	157.0
1985	143.3	162.0
1986	133.0	158.0
1987	113.7	153.5
1988	110.2	149.3
1989	108.3	152.0
1990	107.8	146.3
1991	105.6	139.7
1992	107.4	140.8
1993	109.8	140.3
1994	117.7	139.3

Source: U.S. Department of Labor, Bureau of Labor Statistics.

Table 5-3. Average Undiscounted AT&T Direct-Dialed Inter-LATA Rates, 1983 and 1993–94
U.S.$ per five-minute call

	Interstate		*Percent change*	*Intrastate*		*Percent change*
Mileage	*12/83*	*3/94*	*(1983–94)*	*1983*	*1993*	*(1983–93)*
0–10	0.96	1.15	+19.8	0.57	0.74	+29.8
11–16	1.28	1.15	−10.2	0.77	0.84	+9.1
17–22	1.28	1.15	−10.2	0.95	0.91	−4.2
23–30	1.60	1.20	−25.0	1.19	1.04	−12.6
31–40	1.60	1.20	−25.0	1.39	1.08	−22.3
41–55	1.60	1.20	−25.0	1.55	1.12	−27.7
56–70	2.05	1.25	−39.0	1.74	1.23	−29.3
71–124	2.05	1.25	−39.0	1.91	1.27	−33.5
125–196	2.14	1.25	−41.6	2.03	1.35	−33.5
197–292	2.14	1.25	−41.6	2.16	1.37	−36.6

Sources: AT&T; Center for Communications Management Information (CCMI).
Note: Intrastate data for thirty-two states only.

longest, a structure that clearly could not be justified by current cost conditions. As a result, the shorter inter-LATA intrastate calls have absorbed substantial rate increases. However, the shorter intrastate rates remain below the interstate rates of similar length, while all calls to locations more than 70 miles away are cheaper in the interstate jurisdiction. The intra-LATA market is examined in greater detail in chapter 6. For our purposes here, we simply note that intrastate rates have fallen somewhat more slowly than interstate rates in multi-LATA states.

Examining the differences in intra-LATA and intrastate inter-LATA rates across states as well as interstate rates should shed some light on the impacts of both competition and regulation.[12] However, the available data are uneven across the three markets. In the interstate market, regulated by the FCC, data on prices and quantities of service (calls and switched minutes) are available, but far from satisfactory. The actual transaction prices and total output for all customers are difficult to measure. Data on AT&T's market share of switched access are available for all years, but there are no data on the amount of traffic carried over dedicated or special access lines. The number of resellers is known. In the intrastate market (intra- and inter-LATA), consumer prices are available for at least ordinary switched calls. However, the growing use of discount plans makes these official tariffs for switched calls less representative of all traffic.

Measures of output are particularly difficult to obtain. More and more traffic is apparently unswitched (private line or virtual private network) or switched only at one end (WATS, 800 calls). Traditional regulatory measures of dial-equipment minutes (DEMs) do not capture all of this output, and their accuracy is increasingly questioned. Recently, an alternative measure, switched-access minutes, has been developed, but only for the interstate jurisdiction. We use both of these measures below.[13]

AT&T's market share for each state for all interstate calls is available through 1989, but not thereafter because of the FCC's shift that year to price caps in regulating AT&T. No nonproprietary data exist on the market shares of individual competitors other than AT&T in the intrastate market. Nor are there data on the RBOC share of intra-LATA markets explicitly; however, we provide an estimate below.

12. We cannot provide a similar analysis for Canada yet because entry into the market occurred too recently.

13. These data are reported by the National Exchange Carrier Association to the FCC. Dial equipment minutes are estimates of total holding time in a local switching center. They include originating and terminating message-telephone service (MTS) and 800 services and outbound WATS services. Switched-access minutes are estimated using access revenues and access rates. See Federal Communications Commission, Common Carrier Bureau, *Monitoring Report,* May 1994, pp. 161–62.

Long-Distance Competition in the United States—The Existing Literature

Does AT&T have market power in the long-distance market today? This critical question is central to much of the current telecommunications public policy debate in the United States. It affects policy choices involving FCC regulation of long-distance rates, the AT&T trial court's decisions on allowing RBOC entry into inter-LATA services, and even legislation now pending before Congress. A growing literature addresses this question, but little consensus has emerged.

Eleven recent studies examine competition in the interstate and intrastate markets. Of these, one study examines the intrastate inter-LATA market, five the intra-LATA market, and six the interstate market (one study examines both interstate and inter-LATA markets).[14] These eleven studies are summarized in appendix A to this chapter and in table 5-4. The data, methodology, and results vary substantially; as a result, so do the policy conclusions. Of the eleven papers, five analyze "data"—that is, they do not use an explicit model. Only three papers use either a structural model or a two-stage modeling process.[15] The remainder of the studies use "reduced-form" equations, which are not complete models based on economic theory but rely instead on an intuitive explanation of the data.

Reduced-form equations can appear somewhat ad hoc because the variables chosen for introduction do not come from a rigorous theoretical specification. Hence, the resulting parameter estimates can be biased if the included variables are correlated with some excluded variables. In addition, without a complete theoretical derivation, causality may be unclear. Thus reduced-form equations show correlations between variables, but not necessarily causality. As an example, suppose that in a regression with AT&T's price on the left-hand side (LHS), AT&T's share is included on the right-

14. These studies are listed in the appendixes to this chapter.

15. David L. Kaserman and others, "Open Entry and Local Telephone Rates: The Economics of Intra-LATA Toll Competition," Auburn University, May 1992; Robert Kaestner and Brenda Kahn, "The Effects of Regulation and Competition on the Price of AT&T Intrastate Telephone Service," *Journal of Regulatory Economics*, vol. 2 (1990), pp. 363–77; Simran K. Kahai, David L. Kaserman, and John W. Mayo, "Is the 'Dominant' Firm Dominant? An Empirical Analysis of AT&T's Market Power," *Journal of Law and Economics* (forthcoming).

Table 5-4. Summary of Existing Studies of Competition in U.S. Long-Distance Markets

Author	*Period*	*Market*	*Methodology*	*Reduced form or structural model*	*Major results*
1. Ward 1993	1986–91 aggregate; 1988–91 monthly state	Interstate	Estimate firm specific demand elasticities	Reduced form	AT&T possesses little market power
2. Taylor and Taylor 1992, 1993	1984–92 aggregate national	Interstate	Measure access cost decrease and AT&T's price reductions	Data	AT&T's price decline has not matched access cost decreases
3. WEFA 1993		Interstate		Macroeconomic/ econometric model	Large potential gains in economic welfare from new entry
4. Hall 1993	1985–92 aggregate national	Interstate	Examine CPI, average revenue per minute estimate	Data	AT&T's average revenue per minute declined by much more than the decline in access costs per minute
5. Kaserman and others 1992		Intra-LATA	Estimate	Structural model	Intra-LATA competition has no impact on the price of local service

6. Taylor, Zarkadas, and Zona 1992	1980–94	Intra-LATA	Estimate degree of adoption of new technologies	Reduced form	Deregulation is associated with a slower adoption of new technologies
7. Tardiff and Taylor 1993	1983–91	Intra-LATA	Estimate impact of regulatory schemes on prices	Reduced form	Some forms of incentive regulation are associated with lower intra-LATA prices
8. Mathios and Rogers 1989	1983–87	Intra-LATA and Inter-LATA (AT&T)	Estimate impact of regulatory schemes on prices	Reduced form	Regulatory flexibility in inter-LATA markets reduces prices; blocking competition in intra-LATA markets raises prices
9. Kaestner and Kahn 1990	1980–86	Intra-LATA	Estimate impact of deregulation and pricing flexibility on prices and AT&T's market share	Two-stage reduced form	Regulatory flexibility and competition reduces prices substantially and AT&T's market share
10. Kahai, Kaserman, and Mayo 1995	1984:3 1993:4	Interstate (200-mile AT&T call)	Estimate supply response of competitors to AT&T	Two-stage structural model	Competitors to AT&T can restrain AT&T market power; AT&T has low price-cost margin
11. MacAvoy 1994	1987–93	Interstate, seven different services	Estimate price-cost margins	Reduced form	Price-cost margins have increased since 1990, inconsistent with declining market share—thus tacit collusion exists

Sources: See appendix A.

hand side (RHS), assuming (but not in a structural model) that competition will reduce AT&T's price and market share. However, AT&T's price may be high in some markets, because costs are high in these markets and competitors choose not to enter them. Therefore, even if entry does not reduce rates, there will be a positive association of rates with AT&T's market share.

Given the paucity of data in many of these markets, reduced-form equations do provide useful information. However, no single existing study is sufficiently robust to be a basis for policy judgments. All of the published analyses must be examined to assess the vigor of long-distance competition.

The most controversial of the recent studies is likely that by Taylor and Taylor.[16] Their basic analysis is simple: They compare AT&T's projected interstate revenue declines with its decline in interstate access costs. The authors suggest that AT&T did not pass on these exogenous reductions in access costs to its customers. The interstate market cannot therefore be classified as fully competitive; under full competition, or in a contestable market, the authors argue that exogenous cost reductions are fully passed on. However, the data used by Taylor and Taylor are weak. The authors simply add up the revenue reductions forecast by AT&T in all tariff or price-cap filings and compare this aggregate revenue reduction to the aggregate of access cost reductions.

To remedy the problems with reported price indexes or Taylor and Taylor's revenue-reduction forecasts, Hall uses data he considers superior and shows that interstate rates fell more than did access costs. However, he does not explain his method.[17] As an alternative, we examine below how total minutes of use respond to changes in access costs and in AT&T's market share.

Four of the five studies of the intrastate intra-LATA market tend to corroborate each other. They suggest limited competition and the uneven (and sometimes perverse) effects of deregulation—prices rise when states allow full pricing flexibility. On the other

16. William E. Taylor and Lester D. Taylor, "Postdivestiture Long-Distance Competition in the United States," *American Economic Review*, vol. 83 (May 1993, *Papers and Proceedings*), pp. 185–90.

17. Robert E. Hall, *Long Distance: Public Benefits from Increased Competition*, study prepared for MCI Telecommunications Corporation, Washington, October 1993.

hand, a study by Kaestner and Kahn suggests significant competitive effects of innovative regulatory schemes and deregulation.[18]

We believe that the published studies of long-distance competition lead to the following general conclusions:

—The intra-LATA toll market does not appear to be intensely competitive, but there is little analysis as to whether competition or regulation is at fault, except in the Kaestner and Kahn study that finds deregulation to be extremely effective in reducing prices, likely through increasing competition. The reasons for the lack of effective competition are unequal access and the other constraints on competitors to the RBOCs.

—The degree of competition in the interstate toll market is unclear. Three studies conclude that there is little competition; three others suggest little market power on behalf of AT&T. The problem is to distinguish between the effects of the number of firms and their market shares and the effects of regulation.

Regulation had been asymmetric (until AT&T's successful court challenge requiring tariff filings for MCI and Sprint), creating impediments to AT&T price reductions by requiring tariff filings for AT&T and not for its rivals. The question remains whether unregulated competition among existing firms would reduce prices further, a question that may be answered now that the FCC has declared AT&T "nondominant" in the interstate jurisdiction.

There are therefore two central issues to resolve. First, given the number of competitors, is regulation maintaining high prices and preventing full competition? Second, given the type and degree of regulation, would an increase in the number of firms lower prices? If regulation is central to maintaining high prices, then increasing the number of competitors will not necessarily lead to the intended effect of reducing prices.[19] Only one of the cited eleven studies, that of the intra-LATA market by Kaestner and Kahn, attempts to distinguish these two effects.

In this chapter and the next, we provide our analysis of the three markets. We begin with the interstate market, where relatively simple analyses suggest that competition has reduced prices. We

18. Kaestner and Kahn, "The Effects of Regulation and Competition," pp. 363–77.

19. Of course, competition has other important advantages—promoting efficiency, introducing new services, and expanding innovation, to name a few.

then turn to an examination of the intrastate inter-LATA market in this chapter, using a variant of the approaches used by WEFA and by Kaestner and Kahn. Our conclusion is that competition *in the inter-LATA market* has been effective in reducing prices; however, the markets are not fully competitive so that further entry would be of real value.[20]

We also find that the effects of various kinds of regulatory reform, such as allowing pricing flexibility, have no benefit without equal access and sometimes lead to perverse results such as higher prices, particularly if long-distance rates are already too high. Thus regulatory reform must go hand in hand with introducing competition, especially in intra-LATA markets where entry barriers to competitors still exist.

Long-Distance Competition, Rates, and Network Use

We refer to the entire market—interstate, intrastate intra-LATA, and intrastate inter-LATA combined—as the *long-distance* market.[21] Any other submarket is referred to on its exact basis (for example, interstate or intrastate intra-LATA or inter-LATA). Estimating the effects of competition in these markets or on long-distance use is complicated by the changes in the industry since the advent of competition. First, as we have noted, competition has generated a plethora of rate plans for residential and business customers that are not captured in the price indexes compiled by the Bureau of Labor Statistics (BLS). Second, large business customers may connect to the network in a variety of ways not measured in conventional statistics of network use. Dedicated local telephone company lines, circuits provided by a variety of competitive local-access carriers, and long-distance companies' own dedicated lines to large customers transport traffic that is not measured by those compiling data on minutes of use for the public switched network. As a result, both use and price are difficult to measure.

20. In chapter 6 we examine the intra-LATA market in detail; again, we find some evidence that competition has reduced prices, but real barriers to competition still exist in many states because of the lack of equal access. Thus, although a significant degree of competition is present, more competition is needed.

21. We do not refer here to markets in their antitrust sense, and no inference is to be taken from these analyses about relevant antitrust markets.

Table 5-5 shows the trends in measured telephone rates (from BLS data), switched minutes of use and DEMs in U.S. long-distance markets, U.S. real GDP (in 1987 dollars), and AT&T's share of the long-distance market for the period 1980–93.[22] The temporal patterns in these rates and in network use are so wildly inconsistent that they cast doubt on the validity of any of these data, particularly for recent years. To show this, we divide the 1980–93 period into two periods, 1980–86 and 1986–93.

The 1980–86 period spans the era of AT&T divestiture. In this period, the data suggest that AT&T lost market share in the total long-distance market at an annual rate of 2.7 percent, while real consumer long-distance rates also declined rather slowly. The real CPI for interstate switched service declined at an annual rate of 3.8 percent, while the real intrastate rate declined at a rate of only 1.0 percent.[23] In spite of these modest price trends and a modest 2.6 percent growth in real GDP, however, interstate network use (as measured by DEMs) expanded at a staggering rate of nearly 12 percent per year, while intrastate network use increased at nearly 9 percent per year.

After 1986 these data show that AT&T's share of total long-distance revenues began to decline much more rapidly. In the 1986–93 period, AT&T lost long-distance (revenue) share at a rate of about 5 percent per year. At the same time, the decline in long-distance rates began to accelerate markedly. Real interstate consumer rates declined at an average rate of 7.4 percent, induced in part by the reductions in carrier access mandated by the FCC as it phased in its subscriber line charges. But real intrastate rates also declined much more rapidly in the 1986–93 period (by 6.3 percent per year, compared with 1.0 percent per year in the previous six years). Apparently, competition and access-charge reductions were reducing toll rates everywhere in the United States.

The curious result of this apparent acceleration in competition and rate declines was that network use growth actually *slowed* markedly, as reported by telephone companies' DEMs. After grow-

22. Dial-equipment minutes (DEMs) include each originating and terminating minute for all switched local and long-distance calls but does not include those minutes of use on dedicated private lines, whether provided by the telephone company or other competitive, "bypass" carriers.

23. Remember that these prices are BLS statistics from the consumer price index and underestimate or ignore the pricing and discount plans. See Hall (1993).

Table 5-5. U.S. Long-Distance Rates, Use, and AT&T Market Share, 1980–93

Year	*Real price index (1980 = 100)*		*Use (DEM) (billions of minutes)*		*Interstate switched access (billions of minutes)*	*GDP (billions of 1987 dollars)*	*AT&T share of long-distance revenues*
	Interstate	*Intrastate*	*Interstate*	*Intrastate*			
1980	100.0	100.0	133	141	n.a.	$3776.3	.965
1981	98.2	93.4	144	151	n.a.	3843.1	.959
1982	101.9	94.0	154	158	n.a.	3760.3	.938
1983	100.8	97.5	169	166	n.a.	3906.6	.910
1984	94.5	98.6	208	198	n.a.	4148.5	.901
1985	87.3	96.0	250	222	161.2	4279.8	.863
1986	79.8	94.2	270	237	183.0	4404.5	.819
1987	65.5	89.1	295	252	215.7	4539.5	.786
1988	60.5	83.0	321	269	244.7	4718.6	.746
1989	55.8	75.6	344	286	277.0	4838.0	.675
1990	51.6	70.4	355	300	307.5	4897.3	.650
1991	49.1	66.1	368	304	328.1	4861.4	.622
1992	47.9	63.0	387	318	350.3	4979.3	.608
1993	47.6	60.7	402	323	371.2	5134.5	.581
Annual % change							
1980–86	−3.8	−1.0	11.8	8.7	n.a.	2.6	−2.7
1986–93	−7.4	−6.3	5.7	4.4	10.1	2.2	−4.9

Sources: Rates: CPI—U.S. Bureau of Labor Statistics; DEM and CCL—National Exchange Carrier Association; AT&T Share—Federal Communications Commission; GDP—U.S. Department of Commerce.

n.a. Not available.

ing at a reported rate of nearly 12 percent in 1980–86, interstate network use grew at an annual rate less than 6 percent in 1986–93 despite continued growth in the economy (as measured by real GDP) and a 7.4 percent annual real decline in measured rates. Similarly, annual intrastate toll use growth declined from 8.7 percent (1980–86) to 4.4 percent (1986–93) despite real intrastate toll price declines that accelerated from just 1.0 percent per year to 6.3 percent per year.

How is it possible that this acceleration in rate declines led to a slowing of network growth? *The answer is undoubtedly that both the price and quantity data are woefully inaccurate.* As we have stressed, the CPI data collected by the BLS do not reflect discount plans offered consumers, nor do the producer price indexes (PPIs). In addition, the network-use data reported as DEMs are now less critical to carriers. The shift to rate caps reduces their role in apportioning the carriers' revenue requirement; therefore, there may be large errors in these data.

A potentially better measure of *interstate* network use is available from carrier reports on the number of minutes of switched "carrier common-line" (CCL) or access service provided to long-distance carriers. Because local Bell operating companies (BOCs) are not permitted to offer interstate (or intrastate inter-LATA) long-distance services, they must offer access services to interexchange carriers for all switched interstate service. The interstate switched common-line minutes (shown in column 6 of table 5-5) have been recorded quarterly by the National Exchange Carriers Association (NECA) since mid-1984. Unfortunately, there are no similar data for *intrastate* switched use, inter- or intra-LATA.

The interstate switched-access data suggest a more even pattern and quicker growth of interstate use. However, it is worth noting that 250 billion interstate DEMs were reported to the FCC in 1985, while interstate switched-access minutes reported to all states totaled only 161 billion. We cannot account for a difference of this magnitude. Growth in interstate switched-access minutes averaged about 9.2 percent from the third quarter of 1984 through 1986 and increased to more than 12 percent for the 1987–90 period, declining again in the 1990–93 period.[24] Between 1986 and 1993,

24. The 1984 datum is not shown in table 5-1 because these data are available only from the third quarter of 1984 to the present.

interstate switched-access minutes increased at a rate of 10.1 percent per year. But even these data do not paint a complete picture of interstate use, because the switched-access minutes fail to capture the leakage of demand from the switched public network. As in the case of the DEM data, access services provided through special access lines, the long-distance carrier itself, or competitive access providers are not included in the switched-access data.

In table 5-6 we provide another set of estimates of interstate domestic *conversation* minutes for AT&T, MCI, and Sprint for the 1985–92 period derived from Hall's 1993 study. These estimates put total interstate use at 112 billion minutes in 1985. Table 5-5 shows 250 billion DEMs in 1985. Because there are two DEMs per conversation minute, one for access and one for termination, the DEM data suggest about 125 billion conversation minutes in 1985.[25] Thus, Hall's estimate is only sightly lower than the 1985 estimate from DEM data. However, the growth rate in the data cited by Hall is more than double the growth rate in DEMs over the next seven years (12.6 percent versus 6.2 percent, respectively) and about 1.5 percentage points greater than the growth in switched-access minutes shown in table 5-5.

Column 8 in table 5-6 provides FCC data for total revenue for all interstate plus intrastate long-distance calling for AT&T, MCI, and Sprint.[26] Hall provides estimates of the average revenue per minute in interstate calling for two of the three companies; we use his data to estimate total interstate minutes for the three major companies (column 4).[27] Multiplying total minutes (column 4) by this average revenue yields an estimate of total interstate revenues (column 6). Subtracting this estimate of interstate revenue from FCC estimates of total revenue (column 8) yields an implicit estimate of intrastate revenues for all carriers (column 9). Column 10 provides FCC data on total intrastate revenues for the LECs. Finally, column 11 provides an estimate for total intrastate revenue for *all* carriers (LECs and interexchange long-distance carriers) by adding columns 9 and 10. This results in estimated intrastate revenues that are 37 percent of all U.S.

25. We ignore the effect of calls that are attempted, but not completed, and the requirements to set up the call. The DEM minutes also include the U.S. end of international calls (3.4 billion minutes in 1985). FCC, *Statistics of Communications Common Carriers,* 1985, p. 20.

26. Ibid.

27. Specifically, we assume that Sprint's revenue per minute is the same as MCI's. We then use Sprint's revenues, as published by the FCC, to estimate its total minutes.

Table 5-6. Estimates of Interstate Minutes and Rates and Intrastate Revenues

	1	2	3	4 = 1 + 2 + 3	5	6 = 4 × 5	7	8	9	10	[11 = 9 + 10]
Year	AT&T interstate minutes (millions)[a]	MCI interstate minutes (millions)[a]	Sprint interstate minutes (millions)[b]	Total interstate minutes (millions)	Interstate revenues per minute (dollars)[a]	Estimated interstate revenues (three carriers) (millions of dollars)	Estimated interstate revenues, all IX carriers (millions of dollars)[c]	Total revenues for all IX carriers (millions of dollars)[d]	Estimated intrastate revenues, all IX carriers (millions of dollars)	Intrastate revenues, LECs	Estimated total intrastate revenues (LECs plus IX carriers)
1985	96,935	9,037	6,028	112,000	$0.3041	$34,059	$36,071	$42,630	$6,559	12,185	18,744
1986	113,736	15,533	10,361	139,630	0.2498	34,880	37,225	44,595	7,370	12,873	20,243
1987	133,269	20,136	13,431	166,836	0.2048	34,168	36,661	44,783	8,122	13,736	21,858
1988	137,668	25,979	17,328	180,975	0.1953	35,344	38,417	47,487	9,070	15,113	24,183
1989	145,067	34,554	23,048	202,669	0.1798	36,440	41,409	51,184	9,775	14,840	24,615
1990	159,742	43,219	28,827	231,788	0.1562	36,205	40,726	52,102	11,376	14,690	26,066
1991	166,586	49,757	33,188	249,531	0.1461	36,456	41,952	54,443	12,491	14,115	26,606
1992	174,496	57,325	38,236	270,057	0.143	38,618	45,062	58,368	13,306	13,615	26,921
1993	n.a.	n.a.	n.a.	286,800	0.1423	40,807	47,505	61,533	14,028	13,717	27,745

Sources: All 1993 data estimated from 1993 FCC data on interexchange revenues and NECA data on growth in switched-access minutes.

n.a. Not available.

a. Robert E. Hall, *Long Distance: Public Benefits from Increased Competition*, study prepared for MCI Telecommunications Corporation, Applied Economics Partners, October 1993, except for 1993.

b. Estimated, assuming MCI minutes and Sprint minutes are in same proportion as revenues.

c. FCC, *Statistics of Communications Common Carriers*, 1992–93, IX Revenues, p. 7.

d. FCC, *Statistics of Communications Common Carriers*, 1993–94, p. 7.

toll revenues, almost exactly the same as AT&T reported before divestiture.[28] This lengthy exercise suggests that Hall's methodology provides a reasonable estimate of total interstate revenues; the only question remaining is whether the estimates of the growth in *minutes* and *rates* are also reasonable. However, even if we were to use the NECA data for the growth of switched-access minutes instead of Hall's estimate, we would still obtain an estimate of the average interstate revenue per minute of only 14.4 cents in 1993. This is substantially below WEFA's 1993 estimate of 18 cents per minute, upon which it partially bases its conclusion that the inter-LATA long-distance market is not competitive.

Did Competition Lower Interstate Rates?

Despite all of these problems, it may be possible to use the switched-access data to determine whether increasing competition has led to lower interstate rates. We can do this indirectly and directly, examining the 1984–93 pattern in both switched-access use and rates as reported by the BLS. Our first test is simple and ad hoc: if interstate rates have been reduced because of competition with AT&T, we should expect output to increase as AT&T loses market share. (This hypothesis assumes, conversely, that if competition did not increase, then AT&T's market share would be uncorrelated with total market expansion.) For simplicity, we assume that real[29] interstate rates (P-INTER) depend on the level of FCC-regulated real access charges (ACCR),[30] AT&T's share of total switched minutes (ATT), and a time trend that picks up the effects of technical progress (TIME). We therefore begin by estimating a simple linear price equation:[31]

28. See Laurits R. Christensen, Dianne Cummings Christensen, and Philip E. Schoech, *Total Factor Productivity in the Bell System, 1947–79,* September 1981, table 1.

29. All nominal price data are deflated by the consumer price index. Note that P-INTER is a price index with 1982–84 = 100.

30. Reported by the Industry Studies Division of the Common Carrier Bureau of the FCC in its periodic *Monitoring Reports.* These are estimated average switched-access rates charged by local-exchange carriers (see table 5–1).

31. This simple model ignores the forces that determine AT&T's market share, which include P-INTER, since high prices may induce entry *or* high prices may suggest high costs that deter entry.

$$\text{(5-1)} \quad \text{P-INTER}_t = a_0 + a_1 \text{ ACCR}_t + a_2 \text{ ATT}_t + a_3 \text{ TIME}_t + u_t$$

for each of t = 1984III . . . 1993IV quarters, where u is a random disturbance term.[32] When equation 5-1 is estimated, the best results are obtained when both ATT and its one-quarter lagged value are included:

$$\text{(5-1}'\text{)} \quad \text{P-INTER} = -0.12 + \underset{(t\,=\,7.79)}{0.34 \text{ ACCR}} + \underset{(t\,=\,1.58)}{0.26 \text{ ATT}} + \underset{(t\,=\,2.16)}{0.31 \text{ ATT}(-1)} + \underset{(t\,=\,2.01)}{0.002 \text{ TIME}}.$$

$$R^2 = 0.996$$
$$\rho = 0.718$$
$$DW = 1.613$$

The coefficient, ρ, is the Cochran-Orcutt first-order correction for serial correlation.

For output, we use the number of terminating switched-access minutes as a proxy for *conversation* minutes, because there is less leakage from switched access into leased lines or the long-distance carriers' own circuits in terminating calls than in originating them. The number of these common-line switched minutes, CCLMIN, responds to changes in the access charge, ACCR; the strength of the economy (real GDP); the share of switches equipped with equal access (EQACC); and a time trend, but not AT&T's share:[33]

$$\text{(5-2)}$$
$$\text{CCLMIN} = 1.27 - \underset{(t\,=\,-1.96)}{0.347 \text{ ACCR}} + \underset{(t\,=\,6.08)}{0.0067 \text{ GDP}} - \underset{(t\,=\,2.05)}{6.27 \text{ EQACC}} + \underset{(t\,=\,34.50)}{0.74 \text{ TIME}} - \underset{(t\,=\,-0.33)}{2.59 \text{ ATT}}.$$

$$R^2 = 0.998$$
$$\rho = 0.875$$
$$DW = 1.877$$

32. The equation was also estimated in loglinear form with similar results.

33. CCLMIN is measured in billions of minutes per quarter; GDP is in 1987 dollars at an annual rate. The period of the regression is only 1986-III through 1993-IV, because terminating access data are not available for earlier quarters.

Even when EQACC is dropped, the value ATT does not assume a statistically significant coefficient. These results suggest that interstate rates, as measured by the CPI, respond immediately to reductions in access charges and improvements in competitors' access conditions, but not to changes in AT&T's market share per se.[34]

Obviously, the data we use for rates (the CPI for interstate long-distance) and output (NECA estimates of terminating access) are likely subject to errors in variables problems.[35] To check for the reasonableness of these data, we estimate a loglinear demand equation 5-3 by two-stage least squares.[36] The results are reasonable, although the estimated price elasticity is somewhat lower than that commonly found for long-distance calling:[37]

$$\text{(5-3)} \quad \text{LCCLMIN} = -5.29 - \underset{(t\,=\,-4.87)}{0.453\ \text{LP-INTER}} + \underset{(t\,=\,1.28)}{0.99\ \text{LGDP}} + \underset{(t\,=\,6.57)}{0.012\ \text{TIME}}$$

$$R^2 = 0.998$$
$$DW = 1.752$$

These results are understandable, however, given the errors-in-variables problem with the CPI price series.

As we discussed previously in this chapter, reduced-form equations such as 5-1 and 5-2 show correlations but not necessarily causality.[38] Thus the causality need not run from AT&T's declining market share because of competition to a reduction in price. As an alternative hypothesis, a reduction in AT&T's market share could be associated with lower prices, which themselves fall because of the effects of competition and/or the reduction in access costs mandated by the FCC (as argued by Taylor and Taylor).

34. The results are slightly different when equation 5-2 is estimated in loglinear form, but EQACC still dominates ATT in this transformation.

35. We have no alternative to the CPI because we require *quarterly* time-series data to obtain sufficient degrees of freedom.

36. The instruments for LP-INTER are LACCR, LATT, LATT(-1), and DRATECAP, a dummy variable equal to zero before the imposition of rate caps on AT&T in 1989 and unity thereafter. The L prefixes reflect natural logarithms.

37. See Lester D. Taylor, *Telecommunications Demand in Theory and Practice* (Dordrecht: Kluwer, 1994). Most studies find that the price-elasticity of demand is fairly low (about -0.2 to -0.4) for shorter toll calls, but substantially higher (-0.7 or more) for longer calls.

38. We are indebted to Lester Taylor for this point.

Using two-stage least squares, we therefore estimated equation 5-1a in conjunction with 5-1, now shown as 5-1″:

$$\text{(5-1a)} \quad ATT = 0.49 + 0.35\ \text{P-INTER} + 0.01\ EQACC - 0.002\ TIME,$$
$$(t = 4.44) \quad (t = 0.36) \quad (t = -3.80)$$

$$R^2 = .980$$
$$\rho = 0.222$$
$$DW = 2.120$$

$$\text{(5-1″)} \quad \text{P-INTER} = 0.16 + 0.12\ ATT + 0.40\ ACCR + 0.0018\ TIME.$$
$$(t = 0.75) \quad (t = 10.66) \quad (t = 1.50)$$

$$R^2 = 0.995$$
$$\rho = 0.659$$
$$DW = 1.771$$

In equation 5-1a, AT&T's share is significantly and positively affected by price, but in equation 5-1″ AT&T's share does not have a significant effect on price. This suggests that AT&T's share of the market may be responding to declining rates in the competitive environment, but AT&T's market share is not a good proxy for the degree of competition in the market—at least, not for the 1984–93 period. It is worth noting that the coefficient on ACCR remains robust in equation 5-1″, reflecting the strong effect that lower access charges have had on interstate rates.

AT&T's Prices and Costs

The debate over whether AT&T continues to enjoy market power in the inter-LATA market has obviously not been settled, and our analysis of the interstate market does not advance knowledge much. Before moving to a more sophisticated empirical analysis of the intrastate inter-LATA market, however, it might be useful to review recent data on AT&T rates, costs, and profit margins to place the debate in perspective. After all, the degree of competition in long-distance markets should have some impact on AT&T's reported performance.

In 1993, AT&T reported that it recorded $26.7 billion in revenues from ordinary MTS services and $4.8 billion in WATS and 800 service revenues.[39] Approximately three-fourths of these revenues ($23 billion) is estimated to derive from the interstate jurisdiction. AT&T accounted for 60.2 percent of all interstate switched-access minutes in 1993. Given that AT&T is likely to use special access or other means to deliver its customers' originating traffic to its switches, we once again use the number of terminating access minutes purchased as the measure of its output. In 1993, NECA reported 211.2 billion minutes of terminating switched minutes of access; AT&T's share of these minutes (and, therefore, its conversation minutes) would be 127.1 billion. Thus AT&T's average revenue per interstate conversation minute in 1993 is estimated at 18.1 cents, the value used by WEFA. Using the Hall methodology (table 5-6), we obtain an estimate of 14.2 cents.[40]

In 1993 interstate switched-access cost 6.75 cents per minute (table 5-1). Given that AT&T uses various less expensive forms of access for large customers, its average access costs were undoubtedly substantially less. AT&T's margin over access charges was therefore about 12 cents per interstate conversation minute. In 1993, AT&T reported that it incurred the following costs in delivering interexchange services:[41]

Plant and operations costs	$6.3 billion
Marketing and customer service	$6.6 billion
General and administrative costs	$4.9 billion

If we allocate three-fourths of these costs to interstate operations, the costs per interstate conversation minute are:

Plant and operations costs	3.7 cents
Marketing and customer service	3.9 cents
General and administrative costs	2.9 cents

39. There were also another $4.2 billion in private network and miscellaneous long-distance service revenues that we ignore because a substantial share of such services does not involve regular switched-access arrangements.

40. The difference between the two estimates depends crucially on the number of conversation minutes. Hall's estimates of conversation minutes are much greater than those obtained by using terminating switched minutes of access. The deductions that follow from analyzing AT&T's reported costs, however, do not depend on the accuracy of the switched-minutes data, because all AT&T revenue and cost data are divided by the same estimate to obtain per-minute magnitudes.

41. All AT&T revenue and cost data are obtained from the FCC, *Statistics of Communications Common Carriers*, 1993–94 edition.

AT&T's net income from operations (after taxes) was $2.385 billion in 1993. If three-fourths of this total is once again attributable to interstate activity, it earned $1.79 billion from interstate services, or 1.4 cents per minute.[42]

Obviously, AT&T's reported net income is a very small share of revenues, but low profit margins do not necessarily mean that rates are competitive. Competition could probably squeeze many of the cost categories, especially marketing and general and administrative (G&A). These costs are now substantially higher than access charges, which are commonly thought to make up the largest share of carriers' costs.

The price-cost margin (Lerner index) may only be estimated if we know the correct value of marginal costs. If these are 1 to 2 cents plus access charges, as MacAvoy and others allege, the Lerner index in 1993 was somewhere between 0.39 [(14.2 − 6.7 − 2)/14.2] and 0.61 [(18.1 − 6.0 − 1)/18.1].[43] However, if the marginal costs of delivering and marketing the services are higher (as AT&T's cost data indicate), the Lerner index is obviously lower. For example, if we take long-run incremental costs including plant and marketing to be 5 cents, the Lerner index could be as low as 0.35.[44] If AT&T's actual costs are assumed to be reflective of an efficient competitor in a competitive market, the Lerner index could even be a paltry 0.08.

Although some may argue that the costs of plant (fiber) and marketing are not traffic sensitive and that 1 cent is the correct long-run incremental cost,[45] pricing at 7.7 cents per minute in the long run would yield AT&T revenue of only about $2 billion over and above access costs. This would leave more than $11 billion of the costs of plant and operations, marketing and customer service,

42. It should be noted that we are ignoring the contribution of other (non-MTS) interstate services in this calculation. As a result, we are probably overstating AT&T's profit margin somewhat.

43. In this calculation, we are assuming that AT&T's access costs in 1993 were between 6.0 and 6.7 cents per conversation minute and its rates were between 14.2 cents and 18.1 cents per minute.

44. A summary of existing studies of the price-cost margin (the Lerner index) is found in Timothy F. Bresnahan, "Empirical Studies of Industries with Market Power," in Schmalensee and Willig (1989).

45. See WEFA Group, *Economic Impact of Eliminating the Line-of-Business Restrictions on the Bell Companies* (Bala Cynwyd, Pa.: July 1993); and Paul W. MacAvoy, "Tacit Collusion by Regulation: Pricing of Interstate Long-Distance Telephone Services," Working Paper Series C 37 (Yale School of Management, November 1994).

and G&A uncovered. Pricing at incremental cost given sunk fixed costs cannot lead to a break-even result. It is therefore unwise to estimate AT&T's marginal costs as simply 1 cent per minute over and above access costs and conclude that prices should fall to this level. What is needed (and unavailable in the interstate market and other jurisdictions) is a complete model of the cost of service, including NTS costs and traffic-sensitive costs.

Modeling the Intrastate Inter-LATA Market

Our analysis of the interstate market here is admittedly simplified. We examined the relationship between prices and minutes of use on the one hand and AT&T's market share and change in access costs on the other. There is apparently a positive relationship between prices and both AT&T's market share and the access charge and a negative one between minutes and the access charge. Though these relationships are consistent with a degree of competition (since in an oligopolistic or dominant-firm market, the loss of market share of the largest firm would not necessarily be associated with price reductions), the direct relationship between AT&T's market share and interstate rates is inconsistent with full competition. In fully competitive markets, market share and prices are not necessarily correlated.

These results suggest a market evolving toward a more competitive structure, but one that may not have been fully competitive throughout the postdivestiture era. Our analysis was rudimentary and likely misspecified, because AT&T's market share is clearly endogenous. Prices may be high where AT&T's market share is high, because costs are high and few enter the market. The correlation between P-INTER and ATT may therefore tell us little about the vigor of competition.

A potentially more productive approach to estimating the degree of competition in the long-distance markets is to examine (in a more systematic fashion than Taylor and Taylor) the effects of the exogenous changes in access costs on prices. Various authors have examined competition in cigarette and gasoline markets, determining the degree to which tax changes (exogenous costs) are marked up by the firm as evidence of its market power.[46] A 1 per-

46. On cigarettes, see Jeremy I. Bulow and Paul Pfeiderer, "A Note on the Effect

cent decrease in exogenous variable costs (such as a tax) is fully passed on in competitive markets but not necessarily fully passed on in those that are less than fully competitive. Cigarette and gasoline markets are used in these analyses because tax changes are exogenous and known, and thus allow the researcher to estimate the amount of tax changes passed through to consumers. Most other costs to the firm are not readily observable.

Despite the dearth of fully accurate data on the U.S. long-distance market, we do have tariffs for intrastate inter-LATA message-telephone service (MTS) rates and on access costs.[47] These access costs are exogenous to AT&T, MCI, Sprint, and other firms in the interstate and inter-LATA markets. Moreover, access costs are predominant in the carriers' cost functions. For AT&T they were equal to 50 percent of revenues in 1987 and fell gradually to 38 percent of revenues in 1993. We can therefore use the data on level and changes in access costs to examine to what degree prices reflect these exogenous costs. Because the RBOCs are the major competitors in the intra-LATA market and do not charge themselves for access in these markets, we do not consider access payments exogenous in the intra-LATA market. As a result, we cannot estimate a model relating prices to access costs in the intra-LATA market.

An appropriate model for estimating the effects of competition in the long-distance interexchange market is as follows:[48]

$$\pi_i = (P_i - A) \cdot Q_i - C(Q_i), \quad (5\text{-}4)$$

where π_i = profits of firm i

P_i = price charged by firm i

A = access costs, per minute

Q_i = minutes supplied by firm i

$C(Q_i)$ = costs of output Q_i, excluding access costs.

of Cost Changes on Prices," *Journal of Political Economy*, vol. 91 (February 1983), pp. 182–85; for gasoline markets, see Melvyn Fuss and Leonard Waverman, "Firm-Specific Demand Elasticities, Conjectural Variations, Extent of Marginal-cost Pricing and Competition in Wholesale and Retail Gasoline Markets"; mimeo, University of Toronto, 1993.

47. These are tariffs filed by the carriers with each state. We use AT&T's daytime tariffed rates in all regressions that follow.

48. This model builds on Kaestner and Kahn (1990) and on unpublished work on the gasoline market by Melvyn Fuss and Leonard Waverman (1993).

(Different time periods and different states are ignored for the moment.)

Profits (π_i) are equal to prices minus per unit exogenous costs multiplied by quantity $(P_i - A) \cdot Q_i$, less other costs of operation. For profit maximization:

$$\frac{\partial \pi_i}{\partial Q_i} = P_i - A + Q_i \frac{\partial P_i}{\partial Q_i} - \frac{\partial C_i}{\partial Q_i} = 0, \tag{5-5}$$

or

$$P_i \left[1 + \frac{Q_i \partial P_i}{P_i \partial Q_i} \right] = \frac{\partial C_i}{\partial Q_i} + A, \tag{5-6}$$

or

$$P_i \left[1 + \frac{1}{\epsilon_i} \right] = MC_i + A, \tag{5-6'}$$

where ϵ_i = firm i demand elasticity

and $MC_i = \partial C_i / Q_i$.

Equation 5-6′ simply states that marginal revenue equals marginal cost.

We can rewrite equation 5-6′ by noting that:

$$\epsilon_i = \frac{\eta}{\alpha_i} \tag{5-7}$$

where η = industry demand elasticity
α_i = conjectural variation elasticity[49]

or

$$P_i \left[1 + \frac{\alpha_i}{\eta} \right] = MC_i + A. \tag{5-8}$$

The firm-specific demand elasticity is made up of two components—the industry demand elasticity and a general term, called the conjectural variation elasticity (α_i). The latter measures this firm's estimate of how other firms in the industry would change

49. There is substantial literature debating the value of a model that subsumes the entire structure of firm rivalry and behavior into one "conjectural variation" parameter. See Bresnahan, "Empirical Studies of Industries with Market Power."

their output with respect to a change in firm *i*'s output. For a competitive industry, α_i is zero; equation 5-6′ thus reduces to price equals marginal cost, because the demand curve facing the individual firm is perfectly horizontal—that is, it is infinitely elastic. For a monopoly, the firm's demand elasticity is the industry demand elasticity; thus equation 5-6′ reduces to the standard monopoly pricing condition. Between these two extremes (infinite firm-specific demand elasticity and industry demand elasticity) lie the firm-specific demand elasticities for oligopolistic industries. Estimating the firm-specific demand elasticity or, equivalently, the firm-specific conjectural variations term therefore provides important information on the degree of competition in the market.

Equation 5-6′ can also be rewritten as

$$P_i/(MC_i + A) = \frac{1}{(1 + 1/\epsilon_i)}. \qquad \text{(5-6″)}$$

The ratio of price to marginal cost, including access (the left-hand side of equation 5-6″), is unity in perfect competition ($p_i = MC_i + A$), and at monopoly equilibrium equals the inverse of the industry demand elasticity (η).

For Cournot behavior (as assumed in the WEFA study):[50]

$$P_i\left[1 + \frac{s_i}{\eta}\right] = MC_i + A, \qquad \text{(5-8a)}$$

where s = the firm's market share.

Rewriting equation 5-8a:

$$P_i = [MC_i + A]\left[\frac{1}{1 + s_i/\eta}\right]. \qquad \text{(5-8b)}$$

Equation 5-8b is similar to that estimated by Kaestner and Kahn. Note that the Cournot model predicts that $\alpha_i = s_i$ (compare 5.8 with 5.8a).

For regulated industries, a more complicated model is needed. It begins by stipulating that positive price-cost margins (see the preceding discussion) exist for *two* reasons, regulation that may prevent the firms in the market from competing, and a noncom-

50. The WEFA Group, *Economic Impact of Eliminating the Line-of-Business Restrictions on the Bell Companies* (Bala Cynwyd, Pa., July 1993), pp. 21–24.

petitive market structure.[51] To incorporate regulation into the analysis we allow the *firm's* demand elasticity ϵ_i in equation 5-6′ to be rewritten

$$\epsilon_i = \frac{\eta}{\alpha(R)}, \quad (5\text{-}9)$$

where R is the effect of regulation on the firm's conjecture (specifically $\alpha'(R) > 0$); the effect of regulation is thus to make the market more collusive.

The price-cost margin can then be attributed to the degree of competition (α) *and* to regulation (R). An "improvement" in regulation, all things being equal, while *holding the number of firms in the industry constant* can reduce α (lower the perceived response of other firms) and thus lower the price-cost margin. A change in the number of firms, *holding regulation* (R) *fixed, may or may not reduce the price-cost margin. That is, increasing the number of players but reducing (or increasing) regulatory constraints may not* increase effective competition, if regulation is a barrier to it.

One needs to determine what percentage of any price-cost margin is due to regulation and what percentage to insufficient competition. The papers surveyed here attempt to do this by introducing regulatory decision variables whose presence (or, in the case of Kaestner and Kahn, the length of time the regulatory regime was in operation) provides evidence of the impact of that regulatory regime on prices, all things being equal. However, this is only half the story, since the degree of competition is not examined. Instead, the regulatory regime is a surrogate for competition. Moreover, it can be a misleading story; without accounting for competition, the analyses can attribute either too great or too small an effect on regulation. Only in Kaestner and Kahn is there an attempt to measure the impact of competition (through the market share of every carrier but AT&T).

Access costs represent a marginal cost to IX carriers and are thus akin to a specific tax per minute. Lowering the access costs is equivalent to lowering taxes. Access costs are also exogenous to the firm. How the firm (say, AT&T) responds to access-cost reduc-

51. There is a third reason not discussed here that has some potential relevance to telecommunications. The existence of scale economies means that prices set at marginal cost (MC) will not allow the firm to break even, and some "markup" over MC is needed.

tions provides estimates of the demand curve—that is, the firm-specific demand elasticity, as perceived by AT&T, and thus produces estimates of the degree of competition in the market.

We assume that all minutes of use are tariffed and measured and then examine AT&T's behavior. Defining γ as $[1/(1 + 1/\epsilon_i)]$,

$$P_A = \gamma(MC_A + A), \tag{5-10}$$

where ϵ_A is AT&T's firm-specific demand elasticity and MC and A are the per minute costs and P_A the rate per minute for AT&T. For AT&T to face a downward sloping demand curve in equation 5-10, $\gamma \geq 1$. Jeremy Bulow and Paul Pfleiderer show that there are three functional forms consistent with $\gamma \geq 1$.

$$\log P = a - b\,Q, \tag{5-11a}$$

$$P = a - b \log Q, \tag{5-11b}$$

$$P = a - b\,Q^{\delta} \qquad 0 < \delta < 1 \tag{5-11c}$$

where a, b, and δ are parameters.[52]

As Bresnahan has pointed out, the ability to estimate α (conjectural variations) is dependent on accurate data—measures of prices, quantities, and costs. Ideally one should estimate *jointly* industry demand and firm behavior through equations such as:[53]

$$Q = d(P, \alpha) \tag{5-12}$$

$$P(1 + \alpha/\eta) = MC + A \tag{5-13}$$

In many cases and in this case of modeling toll competition in the United States, we do not have sufficient data to estimate equation 5-12 and therefore use equation 5-13 along with exogenous estimates of the industry demand elasticity (η) to measure α (as is done in WEFA).

A simple approach to measure the degree of competition is to concentrate on γ in equation 5-10, which is the degree to which price is affected by a change in exogenous access costs. When γ equals one, this is consistent with competitive behavior; when a γ

52. Jeremy I. Bulow and Paul Pfleiderer, "A Note on the Effect of Cost Changes on Prices," *Journal of Political Economy*, vol. 91 (February 1983), pp. 182–85. Fuss and Waverman, "Firm-Specific Demand Elasticities." The authors test for the appropriate functional form but use all three equations (5-11a, b, and c) to test for robustness.

53. Bresnahan, "Empirical Studies."

is different from one, this suggests less than fully competitive behavior. In a competitive market, if access costs fall by 1 cent per minute, the price should fall by 1 cent, no less. If access costs go up by 1 cent per minute, price should rise by 1 cent. If γ is well estimated statistically, the exact firm-specific demand elasticity can be calculated; given η (the industry demand elasticity), the firm's conjectural variations term itself can be estimated.

Taylor and Taylor compute the aggregate decline in access costs for AT&T and the predicted decline in aggregate tariffed-service revenues. Over the period 1984–90, they found that the fall in this aggregate revenue was less than the reduction in access costs. But they conclude that if all rate reductions, including discount plans, were included (admittedly a difficult task), the reduction in rates could nearly equal the reduction in access costs.[54] Taylor and Taylor conclude that the market is not competitive and "the explanation for this noncompetitive price behavior is not difficult to find. The seven regional (former) Bell holding companies are barred from the market . . . the Federal Communications Commission instituted a number of measures to protect new competitors . . . and asymmetric regulation of AT&T, which continues to this day."[55]

Estimating the Model for the Inter-LATA Market

The problem with analyzing quarterly or annual national *interstate* data is that there is only one regulator and only eleven years of experience since 1984. The *intrastate inter-LATA* market incorporates thirty-eight jurisdictions and is therefore a promising arena for analysis of the effects of regulation and the degree of competition. Because the MFJ prohibits RBOCs from providing service in this market, access charges are exogenous to the various market participants.[56] Unfortunately, there are no direct data on market shares of participants in the inter-LATA market. We therefore assume these market shares to be the same as in the interstate market, because a carrier with points of presence in a state that offers intrastate inter-LATA service has the same opportunities to offer interstate service to its customers.

54. Taylor and Taylor, "Postdivestiture Long-Distance Competition," p. 180, n. 3.
55. Ibid., p. 187.
56. There is a minor exception to this statement that we deal with below.

We have gathered data on AT&T's undiscounted inter-LATA message-telephone service (MTS) prices by state for a five-minute call for each of eleven mileage bands and the intrastate carrier access rates charged by the BOCs for the years 1987–93. The inter-LATA rates are the published rates paid by residential subscribers and small businesses.[57] Although there are thirty-eight multi-LATA states, we confine our attention to between twenty-nine and thirty-three states in each of our seven years because of missing price data. We also gathered state data on the form of regulation,[58] per capita income (PI/CAP), percentage of population in rural areas of the state (RURAL), percentage of main lines provided to business customers (BUS), population density (DENSITY), and state access charges.[59]

Preliminary Observations

Before estimating effects of variables, we provide a brief review of the data. The top half of table 5-7 provides data on the average *real* price of a five-minute inter-LATA call for eleven mileage bands and four years (1984, 1987, 1991, and 1993).[60] The bottom half of table 5-7 provides similar information except for the price net of access costs. The degree of "rate rebalancing" is readily discerned in this tabulation, as prices in 1993 in the longest mileage bands (7 through 11) are about half of 1984 levels. The bottom half of table 5-7 shows how the margins over access costs have changed over these nine years: the longer the mileage bands, the higher the

57. Over time, a larger share of customers have used various discount plans. Unfortunately, we do not have data on the average price paid after such discounts nor the number of customers choosing discount plans across our seven-year, twenty-nine- to thirty-three–state sample.

58. We characterize regulation as falling into three different regimes: full rate regulation, partial rate flexibility (such as banded, ceiling, or floor regulation), and full flexibility. Therefore, we include two dummy variables to reflect these regimes—FFLEX for full pricing flexibility and NOFLEX for no flexibility. Information on regulatory regimes across states is drawn from the *State Telephone Regulation Report* and a variety of AT&T tabulations.

59. The income data are obtained from the Bureau of Economic Analysis of the U.S. Department of Commerce. RURAL and DENSITY are derived from the Census of Population and the NTIA's *Directory of Local Exchange Carriers*. BUS is taken from the FCC's *Statistics of Communications Common Carriers*. Access charges were supplied by a large telecommunications carrier.

60. Henceforth, our results and data will be shown in 1986 U.S. dollars, using the CPI-U as the deflator.

Table 5-7. Average Daytime Residential Intrastate Inter-LATA Rates
1986 U.S.$ per five-minute call

	Price			
Mileage band	*1984 N=30*	*1987 N=29*	*1991 N=32*	*1993 N=30*
0–10	0.70	0.77	0.59	0.59
11–16	0.92	0.92	0.69	0.66
17–22	1.10	1.02	0.75	0.72
23–30	1.36	1.21	0.86	0.82
31–40	1.55	1.31	0.90	0.85
41–55	1.71	1.38	0.92	0.88
56–70	1.92	1.53	1.00	0.97
71–124	2.08	1.60	1.04	1.00
125–196	2.24	1.71	1.13	1.06
197–292	2.33	1.76	1.15	1.08
293+	2.41	1.82	1.18	1.11
	Price minus access costs			
Mileage band	*1984 N=30*	*1987 N=29*	*1991 N=32*	*1993 N=30*
0–10	n.a.	0.01	0.17	0.23
11–16	n.a.	0.15	0.27	0.31
17–22	n.a.	0.26	0.33	0.36
23–30	n.a.	0.44	0.44	0.47
31–40	n.a.	0.55	0.48	0.50
41–55	n.a.	0.62	0.50	0.53
56–70	n.a.	0.77	0.59	0.61
71–124	n.a.	0.84	0.62	0.64
125–196	n.a.	0.95	0.71	0.71
197–292	n.a.	1.00	0.73	0.72
293+	n.a.	1.06	0.77	0.75

n.a.: Access charge data not available for 1984.

margin over access costs. In 1987, the longest mileage band earned a margin 100 times that of the lowest, but the differential narrowed to 4.5 in 1991 and just 3.3 in 1993. The cost of a 200-mile call (mileage band 10) is not likely to be twice as high as the cost of a 20-mile call (mileage band 3); hence, even in 1993, these rates suggest some price discrimination. Because demand elasticities rise with distance,[61] this type of price discrimination (if it does exist), is inconsistent with Ramsey pricing.

Table 5-8 provides similar data disaggregated by regulatory regime. For each year, average prices are given for states with full rate flexibility, partial rate flexibility, and no price flexibility, as is

61. See Taylor (1994).

the number of states in each regime. In 1987 the average prices in the higher mileage bands were greater in states with partial flexibility than in those with full flexibility. This result reversed in 1991. However, the number of states in each category rose over these few years, and several states migrated across regimes; comparisons of this sort are therefore risky. Regression analysis is required to determine the correct associations.

Changes in the dispersion of prices is also evidence of more aggressive competition. Although we have not reported it, the interstate variability in price declines appreciably over time. For all mileage bands, the standard deviation of price in 1993 (50 percent of the mean) was substantially lower than it was just 6 years earlier (71 percent of the mean). This is particularly true for the shorter mileage bands.

The average intrastate access charge per conversation minute (in 1986 dollars) has also decreased monotonically over time, as has the standard deviation of the access charge averaged over all multi-LATA states.[62] In 1991 the standard deviation of the access charge was 54 percent of its 1987 level and by 1993 had declined to 50 percent of its 1987 magnitude.

We believe these changes reflect a substantial modicum of competition. The dispersion of prices across states and mileage bands within a state is narrowing and reflects competitive pressures.

Table 5-7 also shows that between 1987 and 1991, the average inter-LATA rate net of access costs rose in the first three mileage bands, but declined in the longest seven mileage bands. Unfortunately, we do not have data for the number of minutes in each mileage band. Between 1991 and 1993, however, only the two largest mileage bands evidenced a decline in gross margins (price net of access costs). This appears consistent with MacAvoy's observation that in the interstate jurisdiction AT&T's margins have increased.

In this analysis, however, we are examining rates in the 38 multi-LATA states, not the single FCC controlled interstate market. In the interstate market, the imposition of price caps in 1989 required AT&T to pass on exogenous reductions in costs such as access costs. In the 38 jurisdictions we are analyzing, there have been

62. The averages over our sample are: 1987, $0.1532; 1988, $0.1254; 1989, $0.1100; 1990, $0.0974; 1991, $0.0856; 1992, $0.0788; 1993, $0.0708.

Table 5-8. Average Daytime Residential Intrastate Inter-LATA Rates by State Regulatory Regime

U.S. 1986 dollars per five minutes

Price	Price flexibility											
	Full				*Partial*				*None*			
Mileage band	*1984* N=1	*1987* N=9	*1991* N=12	*1993* N=12	*1984* N=7	*1987* N=17	*1991* N=20	*1993* N=17	*1984* N=20	*1987* N=3	*1991* N=0	*1993* N=1
0–10	0.71	0.79	0.63	0.60	0.83	0.77	0.56	0.58	0.66	0.75	...	0.56
11–16	0.95	0.94	0.72	0.65	1.03	0.93	0.66	0.67	0.89	0.75	...	0.65
17–22	1.12	0.96	0.76	0.70	1.22	1.05	0.74	0.73	1.06	1.00	...	0.65
23–30	1.36	1.14	0.89	0.80	1.46	1.26	0.83	0.82	1.33	1.10	...	1.04
31–40	1.46	1.18	0.92	0.84	1.69	1.39	0.89	0.85	1.51	1.25	...	1.04
41–55	1.57	1.24	0.95	0.87	1.85	1.46	0.91	0.88	1.67	1.33	...	1.04
56–70	1.69	1.44	1.05	0.98	2.08	1.58	0.98	0.95	1.88	1.51	...	1.16
71–124	1.80	1.48	1.08	1.02	2.21	1.66	1.02	0.98	2.04	1.61	...	1.16
125–196	1.89	1.63	1.19	1.09	2.35	1.75	1.09	1.04	2.22	1.73	...	1.24
197–292	1.97	1.67	1.22	1.10	2.43	1.80	1.11	1.05	2.32	1.79	...	1.24
293+	1.97	1.78	1.26	1.14	2.45	1.85	1.15	1.07	2.42	1.79	...	1.35

Table 5-8. (continued)

U.S. 1986 dollars per five minutes

Price minus acess cost	*Price flexibility*											
	Full				*Partial*				*None*			
Mileage band	*1984 N=1*	*1987 N=9*	*1991 N=12*	*1993 N=12*	*1984 N=7*	*1987 N=17*	*1991 N=20*	*1993 N=17*	*1984 N=20*	*1987 N=3*	*1991 N=0*	*1993 N=1*
0–10	n.a.	0.11	0.26	0.22	n.a.	−0.06	0.12	0.24	n.a.	0.05	...	0.27
11–16	n.a.	0.26	0.35	0.28	n.a.	0.11	0.22	0.33	n.a.	0.05	...	0.37
17–22	n.a.	0.28	0.39	0.32	n.a.	0.23	0.29	0.39	n.a.	0.30	...	0.37
23–30	n.a.	0.46	0.52	0.42	n.a.	0.44	0.39	0.48	n.a.	0.39	...	0.75
31–40	n.a.	0.50	0.54	0.46	n.a.	0.57	0.45	0.51	n.a.	0.55	...	0.75
41–55	n.a.	0.56	0.57	0.49	n.a.	0.64	0.46	0.54	n.a.	0.63	...	0.75
56–70	n.a.	0.76	0.67	0.60	n.a.	0.76	0.54	0.61	n.a.	0.80	...	0.87
71–124	n.a.	0.80	0.70	0.64	n.a.	0.84	0.58	0.64	n.a.	0.91	...	0.87
125–196	n.a.	0.95	0.81	0.71	n.a.	0.93	0.65	0.70	n.a.	1.03	...	0.95
197–292	n.a.	0.99	0.84	0.73	n.a.	0.98	0.67	0.71	n.a.	1.09	...	0.95
293 +	n.a.	1.10	0.88	0.76	n.a.	1.03	0.71	0.73	n.a.	1.09	...	1.06

n.a. Access charge data not available for 1984.

different forms of regulation other than simple price-caps. Thus the data provided in table 5-8, showing the rates and gross margins disaggregated by regulatory regime, are crucial. In states with full rate flexibility, price net of access costs was *lower* in 1993 than in 1991 across all mileage bands, but in those with only partial rate flexibility, price net of access costs was higher in 1993 than in 1991. Thus, in the intrastate inter-LATA market, the interaction between regulation and competition (as in equation 5-9) appears clear. Nevertheless, the states change regulatory regimes over time so that the simple results of table 5-8 require further analysis to determine causation.

The distribution of regulatory regimes over our sample period is as follows:

Year	*No pricing flexibility*	*Partial pricing flexibility*	*Full pricing flexibility*	*Total*
1987	3	17	9	29
1988	3	19	10	32
1989	2	17	11	30
1990	2	20	11	33
1991	0	20	12	32
1992	1	20	9	30
1993	1	17	12	30

Unfortunately, there are two missing data. First, as indicated earlier, we have no data on the costs of providing long distance telephone service. We use state and time dummies as well as population density, the percentage of rural population and the percentage of business revenue as proxies for these costs. Second, we have no estimate of output; thus, we cannot estimate either a demand function or the "average" mark-up across all mileage bands to determine whether the mark-up has "on average" fallen or risen. The missing output series is a serious problem that prevents us from developing a full model of the inter-LATA market.

Estimation of the Model

Turning to more structural analyses of competition and regulation, we begin by using the entire data set to regress the prices on a constant, the access charges, time, and a distance dummy (for ten mileage bands):

$$(5\text{-}12) \quad P_A = \underset{(t\,=\,16.4)}{0.465} + \underset{(t\,=\,20.4)}{0.647\ \text{ACC}} - \underset{(t\,=\,-12.5)}{0.04\ \text{TIME}} + \underset{(t\,=\,48.3)}{0.079\ \text{DIST}_i}$$

$$\overline{R}^2 = .62$$

where P_A = price of a five-minute AT&T inter-LATA call
ACC = costs of access (two ends) for a five-minute call
TIME = time dummy 1987 = 1 . . . 1993 = 7
DIST_i = distance band dummy (2–11).

This regression assumes a uniform impact of time and of the access charge across all mileage bands, assumptions that we will show are not necessarily valid. However, certain details stand out:

—The coefficient on the access charge is .65, significantly less than 1. We shall show that this is because of pooling across all mileage bands.

—Prices on average increase 8 cents per mileage band.

—Prices on average fall 4.0 cents per year *after* controlling for access costs and distance.

—The three variables—access charge, time, and mileage band—explain 62 percent of the variation in all prices.

Prices fall by far more than can be explained by access cost changes, suggesting the presence of increasing competition or rapid technical change, or both.

A regression for the price net of access charges across all mileage bands shows the following:

$$(5\text{-}13) \quad P_A - \text{ACC} = \underset{(t\,=\,13.1)}{0.192} - \underset{(t\,=\,-6.8)}{0.017\ \text{TIME}} + \underset{(t\,=\,47.1)}{0.074\ \text{DIST.}}$$

$$\overline{R}^2 = .49$$

In this variant, the price net of access, on average, falls by 1.7 cents per year.

It is far from obvious that our price equation should be the same across all mileage bands. After examining the data, we determined that the mileage bands could be grouped into three categories: 1 and 2, 3 to 6, and 7 to 11.[63] Equation 5-12 was reestimated over these mileage bands:

63. In future variants, we disaggregate the longer mileage bands into 7–8 and 9–11.

(5-12′) $P_A = b_0 + b_1\text{ACC} + b_2\text{TIME} + b_3\,\text{DIST}_i + u$

Mileage Bands:					$\overline{R}^2$
1, 2	0.663	0.042	−0.034	0.11	0.16
	(t = 10.2)	(t = 0.63)	(t = −5.0)	(t = 5.4)	
3–6	0.515	0.531	−0.037	0.077	0.44
	(t = 10.63)	(t = 11.4)	(t = −7.8)	(t = 12.2)	
7–11	0.482	0.982	−0.045	0.055	0.57
	(t = 8.0)	(t = 20.9)	(t = −9.5)	(t = 10.9)	

Our principal concern is measuring the degree of competition in the inter-LATA market. As a result, we are particularly interested in the size of the estimated coefficient of the access charge b_1. It is clear that coefficient b_1 is significantly different across mileage bands: 0 for mileage bands 1 and 2, 0.531 (at the mean) for mileage bands 3–6, and 0.982 (not significantly different from 1.0) for mileage bands 7–11. These initial results suggest significant rate rebalancing.

We now turn to an analysis of AT&T's pricing that includes regulatory variables, state characteristics, and a term reflecting the hypothesis that the marginal cost of an additional minute of calling is lower when market share is higher. Brandon[64] has shown that if regulators adjust the NTS component of access costs by spreading this essentially fixed cost over all long-distance minutes, then AT&T's marginal NTS access cost per minute is zero when it has 100 percent of the market. This rises with its share of the market by a factor equal to $1 - S_A$, where S_A is AT&T's market share. As AT&T's market share declines, the marginal cost of its access minutes rises. To capture this effect, in one variant we divide the access charge (ACC) into its reported NTS component (CCL) and its traffic-sensitive component (ACC-CCL) and multiply CCL by $1 - S_A$.

Because we are estimating a reduced-form equation of the price of long-distance calling, we must include other variables that shift demand and cost across states and over time to produce our final estimates:

64. Paul Brandon is an economist at Bellcore. See Steven C. Parsons and Michael R. Ward, "Telecommunications and the 'Brandon' Effect," Federal Trade Commission, Bureau of Economics Working Paper, February 1993.

(5-14) $P_{Aijt} =$

$$\beta_0 + \beta_1 \text{ACC}_{jt} + \beta_2 \text{TIME}_t + \beta_3 \text{DIST}_i + \beta_4 \text{NOFLEX}_{jt} + \beta_5 \text{FFLEX}_{jt} + \beta_6 \text{RORREG}_{jt} + \beta_7 \text{DENSITY}_{jt} + \beta_8 \text{RURAL}_j + \beta_9 \text{BUS}_{jt} + \beta_{10} \text{PI/CAP}_{jt} + \beta a_{11} \text{ACC}_{jt}{*}\text{NOFLEX}_{jt} + \beta_{12} \text{ACC}_{jt}{*}\text{FFLEX}_{jt} + \beta_{13} \text{ACC}_{jt}{*}\text{RORREG}_{jt} + u,$$

where TIME, DIST, DENSITY, RURAL, BUS, and PI/CAP are as previously defined. TIME, DIST, and DENSITY should capture differences in the marginal cost of calls across time, mileage bands, and states. RURAL, BUS, and PI/CAP are assumed to capture demand shifts across states and time. P_{Aijt} is ATT's price for a five-minute inter-LATA call at mileage band *i*, state *j*, and time *t*. ACC_{jt} is the access charge state *j* at time *t*.[65] The FFLEX, NOFLEX, and RORREG variables reflect states in which there is full rate flexibility, no rate flexibility, and rate-of-return regulation, respectively. In addition, in some variants we substitute $\beta_1^1(\text{ACC-CCL}) + \beta_1^2\text{CCL}(1- S_A)$ for $\beta_1\text{ACC}$.

We begin without including the regulatory interaction terms (ACC*NOFLEX, ACC*FFLEX, ACC*RORREG) and then add the regulatory interaction terms. (The full results are reported in appendix B to this chapter.) A summary of the estimates of the access-charge terms—with and without regulatory interaction terms—are shown in table 5-9. The results are shown for all mileage bands and for mileage bands 1 and 2, 3–6, 7–8, and 9–11 separately.

Disaggregating results across mileage bands is clearly important. The following results stand out:

—Without the regulatory interaction effects, the estimated coefficient β_1 is significantly less than 1, but it rises sharply in the longer mileage bands. In the five longest mileage bands, the estimate of β_1 is between 0.6 and 0.7. The estimates are somewhat higher for just the 1991–93 period.[66]

—The weighted sums of the estimates of β_1^1 and β_1^2 are much greater than the estimates of β_1. For mileage bands 3–6, they sum to 0.456; for the longest bands (7 and 8 and 9–11), they sum to 1.2.

65. We also experimented with a DIST variable that was equal to the average mileage in each band, but the estimates of the remaining coefficients were not affected by this change. Therefore, we do not report these results.

66. For details, see appendix B.

Table 5-9. Estimates of the Degree of Pass-Through of Access-Cost Changes for AT&T's Daytime Intrastate Inter-LATA Rates, 1987–93

	Without regulatory interaction effects			
Mileage bands	β_1 (*All years*)	$[0.57\beta_1^1 + 0.43\beta_1^2]$ (*All years*)	β_1 (*1991–93*)	$[0.57\beta_1^1 + 0.43\beta_1^2]$ (*1991–93*)
1–2	−0.202	−0.311[a]	0.024[a]	0.036[a]
3–6	0.233	0.456[a]	0.291	0.489
7–8	0.614	1.155	0.661	1.119
9–11	0.686	1.242	0.719	1.226
All	0.349	0.656	0.427	0.724

a. Not statistically significant at 5 percent level of confidence.

	With regulatory interaction effects (all years)		
Mileage bands	*With no rate flexibility*	*With full rate flexibility*	*With rate-of-return regulation*
1–2	0.301	−0.339[b]	−0.074[b]
3–6	0.303[b]	0.208[b]	0.278[b]
7–8	0.250[b]	0.899	0.474
9–11	0.021	1.171	0.459
All	0.216[b]	0.497	0.296

b. Regulatory-interaction coefficient not statistically significant at 5 percent level of confidence.

In addition, the complete estimates of equation 5-14 reveal the following (see appendix B for details):

—Prices fall over time after all other effects are accounted for. This decline is probably the result of a combination of increasing competition and rapid technical progress, but we have no good measure of the distribution of competitors' market shares in each inter-LATA market over the period.

—Relative to states with partial rate flexibility, having no flexibility *lowers* rates in the shorter mileage bands, but not in the longer ones. We suspect that this reflects the regulatory pressure to keep short calls priced very low. (We know that competition has led to higher short-mileage rates and lower long-mileage rates.)

—Rate-of-return regulation lowers all rates by 4.2 to 6.2 cents—a somewhat surprising result.

—Rates decline with increases in density, particularly in the longest mileage bands.

—Rates, on average, are higher in the more rural states.

—Rates, on average, are lower in states with a greater percentage of business users.

—States with higher per capita income have generally lower inter-LATA rates, on average, though not in the longest bands.

—The results for the shortest mileage bands are clearly the least precise. This may be because rates in these bands were below competitive-equilibrium rates before deregulation by the states. (Rates net of access charges in the shortest band were near zero in 1987.)

The inclusion of variables that provide for interaction between regulation and the access rate provides at least some hint of how regulation might have impeded competition. The bottom half of table 5-9 shows the sum of β_1 and the various coefficients for the interactive terms β_{11}, β_{12}, and β_{13}. In general, regulatory flexibility increases the degree of pass-through of access costs, whereas no flexibility reduces it. For example, the degree of pass-through increases from 0.614 to 0.899 in mileage bands 7 and 8 when there is full rate flexibility and from 0.686 to 1.171 in the longest mileage bands. No regulatory flexibility sharply lowers the degree of pass-through in the longest bands but raises it in the shorter bands; this suggests, unremarkably, that regulation leans against rate rebalancing. Finally, rate-of-return regulation also impedes the pass-through of access-cost changes in the longer bands.

Is the Inter-LATA Market Competitive?

For our purposes, the question to be answered is the *average* size of β_1, the coefficient of the access charge.[67] Unfortunately, there are no available data on the minutes of use by mileage band by state for the inter-LATA market. Without these data we cannot weight prices by mileage band to arrive at an *average* price. We therefore cannot know the degree to which AT&T passed on access cost differences or decreases, on average, across all mileage bands. Nor can we estimate the demand function; therefore, we offer no direct estimates of η, the price elasticity of demand.[68]

That β_1 is lower for small mileage bands suggests there was significant rate rebalancing by AT&T during the period under study. The mean estimate of β_1 (table 5-9) in mileage bands 7 and 8 is 0.614, and the weighted average of β_1^1 and β_1^2 is 1.155. For the longest bands, the estimates are 0.686 and 1.242, respectively. For

67. Or, alternatively, the weighted average of $\beta_1^1 + \beta_1^2$. In the notation of (5-10), γ; or b in (5-12′).

68. We tested both the linear and loglinear specifications, but discovered no significant differences in the important parameters. Therefore, our results are consistent with an industry demand curve of constant elasticity.

all calls to destinations more than 55 miles away, the estimates are above 0.6 and perhaps around unity.[69] The degree of pass-through is much lower in the shorter mileage bands, which may reflect an attempt by the carriers (AT&T, in this case) to spread out rate rebalancing over time, thus making it more politically acceptable.[70]

In our model, β_1 is equal to $[1/(1 + \alpha/\eta)]$, where η is the market elasticity of demand and α is the conjectural variations term for AT&T. The consensus estimate of the market elasticity of demand for long-distance services is about -0.7.[71] Such an elasticity of demand is obviously inconsistent with the common belief that AT&T has considerable market power in long-distance services, for no firm with market power would operate in the inelastic range of its demand curve. For Cournot oligopoly, however, equilibrium in the inelastic range of the demand curve is feasible as long as the oligopolist's market share is less than the absolute value of the elasticity of demand. Obviously, an AT&T market share of 0.9 (circa 1985) and a demand elasticity of -0.7 do not meet this test, but AT&T's recent market share of about 0.6 and a market elasticity of -0.7 provide a feasible Cournot equilibrium.

In our model, the assumption of an "overall" market elasticity of -0.7 poses several problems. First, there is substantial evidence that the demand elasticity increases with distance.[72] Second, it is quite likely that as the market for long-distance services expands, the market elasticity of demand for any mileage band changes, particularly as competition increases. Our model implicitly assumes a constant demand elasticity, but our inability to muster the output data to estimate this elasticity is clearly a handicap.

Given that AT&T's market share has now fallen to about 0.6 and was between 0.6 and 0.8 for our period of estimation, the elasticity of demand, η, must be greater in absolute value than $-.8$ to provide a feasible Cournot equilibrium. Given the rebalancing of rates that was taking place, it is likely that our model provides estimates of

69. We hedge on the latter assertion. Our attempt to estimate the Brandon effect relies on an arbitrary assumption that the CCL data are correct and that state regulators set access charges by spreading CCL costs over total switched-access minutes.

70. Melvyn Fuss has suggested to us an alternative theory. A monopolist constrained by rate caps would rebalance rates by raising rates in the least price-elastic markets. However, AT&T was *not* faced with rate caps in most of its intrastate markets during our sample period.

71. See L. Taylor (1994), chapter 6.

72. Ibid.

β_1 that are biased downward for the shorter mileage bands. In the longer bands, however, the demand elasticity is likely to be greater than -0.7 and perhaps as great as -1.0. With AT&T's market share of 0.6 and a demand elasticity of -1.0, the Cournot equilibrium value of β is 2.5; therefore, an estimate of β less than 2.5 suggests competition that is *more* vigorous than Cournot. Since most of our estimates of β_1 cluster around 1.0, particularly in the longer mileage bands, we conclude that inter-LATA competition is more vigorous than that predicted by the Cournot model.

Recall that in our model, α reflects the interaction of the effects of competition and regulation. The results in tables 5-8 and 5-9 demonstrate that the time profile of inter-LATA rates net of access costs varies considerably across regulatory regimes. With full rate flexibility (20 states), rates net of access charges fall between 1991 and 1993 while they rise in the twelve states with only partial flexibility. Regulation matters. Finally, rates fall over time after all other influences, including access charges, are accounted for (equations 5-12 and 5-14 and appendix B). The combination of all of these results suggests to us that considerable competition exists in the interstate inter-LATA market, but not perfect competition. It is also clear that the continuation of regulation impedes competition.

Without knowing the minutes of service in each mileage band, a stronger conclusion cannot be drawn. However, the size of estimates of β_1 in the longer mileage bands, the decline in rates over and above access cost changes, and the decrease in dispersion over time (both within a state and among states), suggest that the market is characterized by considerable competition.

International Rates

International rates are governed by the settlement rates negotiated by the (typically national) carriers, because each carrier must pay one-half of the settlement rate to the other country's carrier to terminate its international call in the foreign country. As competition among long-distance carriers develops, these settlement rates might be expected to decline and therefore, the price of an international call would as well. But rates between the United States and Canada *in either direction* should be priced similarly if competition exists on both sides of the border. In fact, AT&T rates

Table 5-10. Long-Distance Telephone (MTS) Cross-Border Rates during Peak Hours, Canada–United States, June 1, 1993

U.S. dollars per minute

Carrier		Message telephone service	PROWATS		Megacom 8	800		2Megacom 800
			$300 (monthly)	*$1,000 (monthly)*	*$10,000 (monthly)*	*$300 (monthly)*	*$1,000 (monthly)*	*$10,000 (monthly)*
United States–Canada								
AT&T	300 mi.	0.45	0.42	0.42	0.36	0.53	0.50	0.44
	2000 mi.	0.68	0.61	0.61	0.54	0.68	0.64	0.50
Canada–United States								
BCTel	300 mi.	0.40	0.26[a]	0.26[a]	0.20[a]	0.27	0.24	0.21
	2000 mi.	0.44	0.26[a]	0.26[a]	0.20[a]	0.31	0.27	0.24
Bell Canada	300 mi.	0.40	0.21[a]	0.21[a]	0.20[a]	0.26	0.24	0.20
	2000 mi.	0.44	0.21[a]	0.21[a]	0.20[a]	0.31	0.27	0.24
AT&T Interstate								
	300 mi.	0.23	0.22	0.22	0.15	0.22	0.20	0.16
	2000 mi.	0.25	0.24	0.24	0.19	0.24	0.22	0.17
Bell Canada Interprovincial								
	300 mi.	0.35	0.23	0.23	0.21	0.32	0.31	0.25
	2000 mi.	0.38	0.23	0.23	0.21	0.34	0.32	0.27

Source: SRCI (CRTC) 31 May 1993—116 RRF, converted from Canadian dollars to U.S. dollars at rate of 1.27.

a. Rates are for WATS.

Table 5-11. International MTS Rates, United States (AT&T) and Canada (Stentor), 1994–95

U.S. dollars per five minutes, peak hours

Peak-hour call destination	*Canada (Stentor)*	*United States (AT&T)*	*Ratio: Canada/United States*
Australia	4.10	8.35	0.49
Austria	4.42	6.97	0.63
Belgium	5.08	7.84	0.65
Brazil	6.13	8.25	0.74
Chile	6.13	7.84	0.78
France	3.32	6.50	0.51
Germany	4.37	6.30	0.69
Israel	7.70	9.22	0.84
Japan	5.59	8.20	0.68
South Korea	6.48	9.22	0.70
South Africa	5.74	7.89	0.73
Turkey	7.70	9.48	0.81

Sources: Canada—Stentor, *National Services Tariff*, June 1994; United States—AT&T, April 1995.
Note: Canadian rates converted to U.S. dollars at $0.71 C/$1.

to Canada at peak hours are much higher than are calls from Canada delivered by Bell Canada or BCTel (table 5-10).

Moreover, the international MTS rates charged by AT&T during peak hours are uniformly higher than those charged by Stentor members (table 5-11). Why these great disparities should exist is frankly a puzzle. Long-distance competition has been around much longer in the United States. We have discovered substantial competition in U.S. intrastate inter-LATA markets. Rates for the longer calls are even lower in the U.S. *interstate* market. Why, then, has competition not driven down ordinary MTS rates from the United States to foreign countries to as low a level as is now found in Canada?

Conclusion

This long journey leads us to conclude that the existing studies of the vigor of competition in U.S. interstate and intrastate inter-LATA markets are not particularly convincing and do not lead to a single conclusion. We have attempted to add to this literature through a new analysis of the intrastate inter-LATA market. In so doing, we conclude that ordinary tariffed MTS rates during peak hours reflect substantial competition in the longer mileage bands. This is not to say that even more competition would not put greater downward pressure on long-distance rates in both the intrastate inter-LATA and interstate markets. Indeed, AT&T currently spends a large fraction of its long-distance revenues on marketing, customer service, and G&A. We would expect these costs to be squeezed somewhat by further entry.

Unfortunately, we have nothing new to offer concerning the vigor of Canadian long-distance competition. Interprovincial rates remain substantially above U.S. interstate rates but are likely to fall as competition from Unitel and Sprint Canada develops. As this book is going to press, Unitel is incurring substantial operating losses—losses severe enough to contribute to the pressure for the CRTC to reexamine its regulatory policy.

Finally, we have discovered that U.S. international rates to Canada are far higher than those available to Canadians calling the United States, despite the longer history of long-distance competition in the United States. This finding remains a mystery.

Appendix A

In this appendix, we review the major studies of long-distance competition referred to in the opening sections of chapter 5. In so doing, we attempt to provide a brief summary of the methodology used, the data used in the analysis, and the empirical results in each study.

(1) Ward[1]

Ward estimates firm-specific demand elasticities in the interstate market (using annual national data, 1986–91; monthly state data, 1988–91) and finds near-perfect competition—that is, little or no market power on the part of AT&T.

METHODOLOGY. Ward uses a two-stage model where aggregate telephone demand is estimated in stage 1 and firm market shares in stage 2. This model assumes that competition between firms affects total demand only through competition's effect on the average price of long-distance services. Ward uses the relationship:

$$n_{ii} = n_{ip} w_i + n_{ii}^c,$$

where n_{ii} = firm-specific demand elasticity

n_{ip} = firm-specific demand elasticity with respect to industry average price (stage 1), holding firms' relative prices constant (stage 2)

n_{ii}^c = firm-specific demand elasticity conditional on industry average price

w^i = firm's market share

Ward estimates the n_{ii}^c for AT&T plus "an aggregation of MCI and Sprint," as follows in a fixed-effects model:

$$w_i = (1 + n_{ii}^c) \ln p^i + n_{ii}^c \ln p^J + \Sigma\lambda \text{ state} + \Sigma \text{ month}.^2$$

RESULTS. The estimated firm-specific elasticities are fairly high. "The deadweight loss due to supra-competitive pricing is . . .

1. Michael Ward, "Market Power in Long Distance Communications," Federal Trade Commission, July 1993.

2. Ward states that the estimate of n_{ii}^c from this equation is biased, because n_{ii}^c is a result of demand and supply equations. Ward accounts for this in the estimation.

estimated to be between half and three-quarters of [1] percent of industry revenue [$300 million to $420 million]. . . . In the intervening decade [since the early 1980s], competitive pressures on AT&T increased substantially."[3]

(2) Taylor and Taylor[4]

Several papers by William and Lester Taylor examine the degree to which interstate toll prices have fallen in the 1972–83 (predivestiture) and 1984–92 (postdivestiture) periods. They find no evidence that competition has reduced interstate toll rates.

METHODOLOGY. The analysis has two components. First, the reduction in interstate toll prices is compared to exogenous decreases in access charges. The authors conclude that "by July 1992, AT&T was receiving annual reductions in access charges (and other exogenous costs) of approximately $10.86 billion. At the same time, AT&T's cumulative price reduction produced only $8.22 billion less revenue per year, compared with 1984."[5] The authors therefore conclude that AT&T's real prices, net of access and exogenous costs, fell twice as fast predivesture as postdivestiture. Second, the authors estimate a demand function for interstate switched services over quarterly data from the third quarter of 1984 through the second quarter of 1992. They then calculate the rate of growth of demand that is unexplained for the 1972–82 and 1989–91 periods. "Growth in demand unexplained by changes in prices, income, and population averaged 1.33 percentage points *lower* in the 1984–91 period than in the 1972–82 period."[6]

There are several problems with these simple analyses of complex issues. First, as the authors realize, estimating the cumulative price reduction is not easy. In Taylor and Taylor, "Postdivestiture Long-Distance Competition," AT&T's price changes are calculated as of each price change date and cumulatively. These price reductions are mainly taken from AT&T's price-caps filings and are "at

3. Ward, "Market Power," p. 33.

4. William E. Taylor and Lester D. Taylor, "Postdivestiture Long-Distance Competition in the United States," *American Economic Review*, vol. 83 (May 1993, *Papers and Proceedings*), pp. 185–90.

5. Ibid., p. 186.

6. Ibid., p. 188.

forecasted demand levels that include stimulation from *anticipated* AT&T rate reductions."[7] That is, the price reductions used by Taylor and Taylor are the ex-ante calculations made by AT&T in support of these price reductions to the FCC. The authors also calculate the "total AT&T price reduction directly from AT&T's actual price index."[8] However, both of these calculations are suspect. Ex-ante calculations that assume demand increases given some elasticity are unreliable. The calculation of price indexes is complicated—increasingly so because of the number and variety of optional calling plans, many of which involve fixed monthly fees. The growing use of optional calling plans[9] is acknowledged but the argument is made that the effect is small (equivalent to a 0.9 percent annual rate of price reduction, based on 1989–91 AT&T estimates). We consider these estimates low. In tables 5-5 and 5-6, we compare the number of switched minutes reported by carriers with Hall's estimates of actual minutes. The difference is large and suggests substantial leakage from switched services if Hall's data are accurate.

(3) WEFA[10]

The WEFA Group uses a macro econometric model, along with a number of assumptions about the degree of competition in the interstate and intra-LATA MTS markets, to estimate the economic effect of allowing RBOC (re)entry into this market. WEFA finds substantial monopoly power in present markets (as of 1992).

METHODOLOGY. The analysis uses a simple Cournot model, among three players of equal size (AT&T, MCI, and Sprint). The study finds high price-cost margins (the opposite of Ward's results), with average interstate prices of $0.18 and average incremental costs of $0.065. Adding a fourth player (the RBOCs) leads to sharply reduced prices and huge consumer benefits: "inter- and intra-LATA toll call rate decreases of 50 percent as compared to Baseline,

7. William E. Taylor, "Effects of Competitive Entry in the U.S. Interstate Toll Markets: An Update" (National Economic Research Associates, Inc., May 1992), exhibit 1, fn. 16.

8. Ibid., exhibit 1, p. 1.

9. Taylor and Taylor, "Postdivestiture Long-Distance Competition," p. 187.

10. The WEFA Group, *Economic Impact of Eliminating the Line-of-Business Restrictions on the Bell Companies* (Bala Cynwyd, Pa., July 1993).

saving consumers $490 billion by 2003."[11] The authors assume that entry by the RBOCs, adding an equivalent-sized competitor, would reduce prices by 50 percent. It is this assumption that is key. These price declines ripple through the economy—stimulating demand, lowering equipment prices, and boosting aggregate U.S. GNP. "The economy gains 3.6 million additional jobs over the next ten years.... $247 billion is added to total real GDP by 2003, a total gain of 3.6 percent over the ten-year interval."[12]

The WEFA analysis, however, compounds two effects: given the industry structure, the effects of regulation may keep prices high; and given regulation, what is the effect of a change in the number of firms from three to four?

The WEFA analysis purports to measure the effect of increasing the number of firms by one while simultaneously removing all regulatory restraints. However, what proportion of the change in GNP attributable by WEFA to entry is in fact the result of entry, and what proportion would be caused by deregulation even if there were no entry? There are potentially both regulatory and competition effects, which should be disaggregated as follows:

—Keeping the number of firms at three, what is the effect of regulatory liberalization?

—Given existing regulatory constraints, what is the effect in increasing the number of players from three to four?

A problem is the use of a Cournot model to examine the interstate long-distance market. The Cournot model assumes three players of equal size—not an appropriate assumption for AT&T, MCI, and Sprint, as evidenced by the data in table 5-4.

Other issues mar the WEFA study[13]—for example, the estimate of marginal costs of interstate toll is $0.01 per minute plus access costs of $0.065. With a measured price of $0.18, the price-cost margin is extremely "high," inconsistent with reported profits that show total costs at least twice as high.

11. Ibid., p. 4.

12. Ibid., p. 3.

13. See Robert E. Hall, *Long Distance: Public Benefits from Increased Competition*, study prepared on behalf of MCI Telecommunications Corp. (Applied Economics Partners, October 1993).

(4) Hall[14]

Robert Hall critiques the Taylor and Taylor paper in a study prepared for MCI. Hall argues that AT&T's prices have fallen more than its access costs. That MCI would attempt to refute the argument that AT&T was not acting competitively appears odd at first glance. However, this study is intended to show that divestiture has worked, competition is alive and well, and the RBOCs should not be allowed to reenter the long-distance interstate market.

Hall begins by examining four measures of the decline in toll prices: CPI, PPI, carrier revenue per minute, and the price-cap index as measured in submissions since 1988 under FCC price-cap regulation. Hall argues that CPI (from BLS data) and PPI do not capture the movements of consumers to new products, because these new services are not included in the baskets. Between 1985 and 1992, average revenue per minute for AT&T, MCI, and Sprint fell 63 percent, while the long-distance component of CPI fell 34 percent and of PPI 24 percent. Taylor and Taylor use the BLS data and AT&T's ex ante calculations of the revenue effects of price reductions in their study. AT&T data filed under the FCC's price-cap regulation show a 15 percent to 17 percent decline in toll rates in the 1988–93 period after accounting for access charges. However, as the price cap was CPI minus productivity (3 percent), a 15 percent to 17 percent decline in toll rates is essentially explained fully by productivity and the price-cap formula, not competition. Table 5A-1 provides data (taken from Hall) in current dollars on decrease in revenue per minute, decline in access costs, and the decline in revenue per minute not resulting from a decline in access costs. These data show that average revenue per minute fell 8.7 cents (from 30.4 cents to 21.7 cents), or 29 percent over the 1985–92 period, after accounting for access-cost decreases. If productivity improvements were 3 percent per year over this seven-year period, productivity increases would have lowered costs and prices (if passed through totally) by 23 percent, or by 7 cents.

Hall also examines entry barriers. They may be high to those attempting to enter on a large-scale, national basis, but low for existing firms and new niche players. Hall estimates AT&T's return

14. Ibid.

Table 5A-1. Changes in AT&T Revenue and Access Costs Per Minute
Current dollars

Year	*Decline in revenue*[a]	*Decline in access cost*[b]	*(1) – (2)*[c]
1985	n.a.	n.a.	n.a.
1986	.054	.009	.045
1987	.045	.020	.025
1988	.010	.010	n.a.
1989	.015	.011	.004
1990	.024	.014	.010
1991	.010	.007	.003
1992	.003	.003	n.a.
Sum 1985–92			.087

n.a. Not available.
a. Hall, *Long Distance: Public Benefits from Increased Competition*, figure 4, column 2.
b. Estimated from Hall using figure 5.
c. Access per minute in 1985 U.S. dollars and Hall, figure 4, column 3, GDP deflator.

on assets (net operating income net of taxes divided by the book value of fixed assets) to be 9.7 percent, not much more than MCI's return of 9.0 percent. If AT&T did not pass through any of its $2.8 billion in access-cost reductions between 1985 and 1992, AT&T's return on assets in 1985 would have been negative, because AT&T's total operating income net of taxes in 1992 was $2.1 billion. In 1985 AT&T reported $1.0 billion in net income.[15]

Table 5A-1 provides data from Hall for the 1985–92 changes in AT&T's revenue per minute and access costs, both in current dollars.[16] The last column in table 5A-1 shows that AT&T's average revenue fell by 8.7 cents more than the decline in its average access costs over this period. Most of this decrease took place in two early years, 1986 and 1987.

15. AT&T Communications results as reported in FCC, *Statistics of Communications Common Carriers*, 1985 and 1992–93.

16. Much of Hall's analysis is undertaken with real data, and he concludes that AT&T's real average interstate price falls substantially. However, after 1984, the residential and small business tariffs for AT&T were regulated under a CPI-3 percent price-caps process; therefore its average revenue from tariffed interstate services must have fallen by 3 percent a year. This was the result of competition only in so far as the price-caps process could not have been instituted earlier, when AT&T's market share was much higher.

(5) Kaserman, Mayo, Blank, and Kahai[17]

Kaserman and others examine intrastate intra-LATA competition using a residual pricing model. For 1990, they find no effect of competition on intra-LATA rates.

METHODOLOGY. The model assumes that the BOCs can set the price of carrier access at the monopoly level and that the price in the intra-LATA toll market results from dominant firm pricing. The price of monopoly local service is the residual—that is, its price is determined by the revenue requirement and profits in the other two markets. The authors show theoretically that increasing competition in the intra-LATA market may or may not increase local rates, because intra-LATA competitors pay access rates to the BOCs. Thus the BOC loss in intra-LATA traffic is offset by access fees and traffic stimulation. The authors use two tests. First, they examine whether the 1990 average price for local service is higher or lower where intra-LATA competition is allowed; they find there are no statistical differences. In a second test, they estimate a variant of the residual pricing model where local telephone rates depend on costs, subsidies flowing from other services, and competition.

RESULTS. No evidence is found that intra-LATA competition affects local prices.

(6) Taylor, Zarkadas, and Zona[18]

Taylor, Zarkadas, and Zona examine the effects of different forms of regulation on the implementation of new technologies in the intra-LATA markets over the 1980–94 period (data for 1990–94 are forecasts). Their econometric results are imprecise, because the variation in rates of diffusion of new technologies is not large. However they conclude that there is no evidence that rate-of-return

17. David L. Kaserman and others, "Open Entry and Local Telephone Rates: The Economics of Intra-LATA Toll Competition," Auburn University, May 1992.

18. William E. Taylor, Charles J. Zarkadas, and J. Douglas Zona, "Incentive Regulation and the Diffusion of New Technology in Telecommunications," paper prepared for the ninth annual conference of the International Telecommunications Society, June 1992.

regulation reduces incentives to innovate. In addition they find that deregulation reduces technological advance. "Slower implementation of these technologies [occurs] when the intra-LATA market is opened to competition."[19]

These results are consistent with our analyses of pricing in the intra-LATA market. We argue that there is indeed *no* effective competition in the intra-LATA market because of the lack of equal access. As a result, allowing flexibility does not necessarily increase competitive pressures. Similarly, Taylor, Zarkadas, and Zona do not show that full competition would necessarily retard innovation.

(7) Tardiff and Taylor[20]

Tardiff and Taylor examine the effects of various forms of regulation at state levels on intra-LATA toll prices for the 1983–91 period for five-minute calls in eight mileage bands.[21] The authors find that some forms of incentive regulation are associated with lower intra-LATA toll prices. (The opposite conclusion is reached by Kaserman, Mayo, Blank and Kahai.)

METHODOLOGY. These authors use a fixed-effects model of form:

$$p_{it} = \alpha_i + \gamma_t + \Sigma B_k \,\mathrm{Reg}_{it} + k_i \,\epsilon_{it}$$

where α_i = state dummy variable
γ_t = time dummy variable
Reg_{it} = regulatory scheme, state i, time t
k = mileage bands

Tardiff and Taylor argue that the fixed-effects model nets out state effects (for example, demographic, political, or geographic forces) that are constant over time and time effects that are constant across companies in a specific period. This is not a fully developed causal

19. Ibid., p. 19.

20. Timothy J. Tardiff and William E. Taylor, "Performance under Alternative Forms of Regulation in the U.S. Telecommunications Industry," study prepared for AGT Limited (Cambridge, Mass.: National Economic Research Associates, April 13, 1993).

21. 0–10 miles, 11–16 miles, 17–22 miles, 23–30 miles, 31–40 miles, 41–55 miles, 56–70 miles, and 71–124 miles.

model based on economic theory. What it shows is correlations between regulatory forms and prices, given two fixed effects.

The authors find that some forms of incentive regulation are associated with lower toll prices (8 percent). They also examine local rates and find no relationship between forms of incentive regulation and these local rates.

In fact, only one form of regulatory reform, *sharing*, is associated with a significant decrease in toll prices. Sharing, according to Tardiff and Taylor, was in place in five states—Florida (1989–92), Idaho (from 1990 on), Kentucky (1988–90), Rhode Island (1989 and 1990), and Wisconsin (1987–89). In two of these states (Kentucky and Wisconsin) sharing was experimental. The Wisconsin experiment was replaced by a revised scheme that does not involve sharing.[22]

Sharing is thus an unstable variable; it changes over time and differs significantly over time within a state. Although we also use dummy variables to represent regulatory reforms, it must be stressed that, in fact, states have quite different regulatory structures even when these are given a common name.

(8) Mathios and Rogers[23]

Alan Mathios and Robert Rogers analyze the effects of regulatory changes on AT&T inter-LATA and BOC intra-LATA rates through 1987. At this early date, relatively few states had a long history of allowing carriers pricing flexibility; none had allowed full competition (including equal access) in intra-LATA markets. For instance, the authors discover that only fifteen states had allowed AT&T any pricing flexibility before 1985.

The authors estimate a pooled regression of the price of a five-minute call in each of ten mileage bands across all states (thirty-nine for the inter-LATA study, forty-eight for the intra-LATA study).

22. The effect on individual regulatory schemes is not shown.

23. Alan D. Mathios and Robert P. Rogers, "The Impact of Alternative Forms of State Regulation of AT&T on Direct-Dial, Long-Distance Telephone Rates," *RAND Journal of Economics*, Autumn 1989, pp. 437–53; Alan D. Mathios and Robert P. Rogers, "The Impact of State Price and Entry Regulation on Intrastate Long-Distance Telephone Rates," Federal Trade Commission, November 1988.

They use a double-log specification of the form:

$$\mathrm{LP}(87) = a_0 + a_1\mathrm{LZ} + a_2R + a_3\mathrm{LP}(83),$$

where LZ is the logarithm of various cost-related and demand variables (such as wage rates, income per capita, density, and rural concentration of population); R is a set of dummy variables reflecting regulatory conditions; and LP(87) and LP(83) are the rates for a five-minute call in 1987 and 1983, respectively. This methodology assumes the same effects of each dependent variable on each mileage band, even though there is substantial evidence that regulatory flexibility would induce the carriers to reduce the spread between the rates for long- and short-distance calls. The inclusion of the 1983 rate allows for state-specific forces that are not included in the authors' analysis that would have influenced rates before divestiture.

In the inter-LATA analysis, the regulatory variable is simply one that equals unity if there is *any* price flexibility allowed to AT&T and equals zero otherwise. Their analysis thus does not distinguish among various degrees of regulatory flexibility; nor does it capture the effects of different degrees of competition in the market. Nevertheless, the coefficient on the regulatory variable is highly significant and negative, suggesting that regulatory flexibility in *inter-LATA* markets leads to lower rates.

The intra-LATA analysis includes regulatory dummy variables for competition (although not with equal access) and the blocking of resale, and the product of the two. Of these, only the competition variable is statistically significant. This indicates that states that block competition in intra-LATA markets (resellers and facilities-based competition) have rates that are about 7.5 percent higher than those that do not.

Our results for intra-LATA rates in chapter 6 use somewhat the same techniques as Mathios and Rogers, but they do not pool across all mileage bands and they are for a much longer and more recent period. We obtain similar but less robust results for the competition variable, but we find that price deregulation actually raises intra-LATA rates. The inter-LATA results in this chapter use a totally different model, but they estimate the effects of regulation by groups of mileage bands, unlike the aggregation across all mileage bands used by Mathios and Rogers.

(9) Kaestner and Kahn[24]

Robert Kaestner and Brenda Kahn analyze developments in the intrastate toll market over the 1986–88 period. The authors attempt to distinguish the effects of an increase in the effectiveness of competition as well as on the regulation of AT&T. They find that deregulation and competition have significant effects on prices.

METHODOLOGY. Kaestner and Kahn begin with a classic conjectural variations model:

$$p = \text{MC}\ \{1\backslash[1 - (S(1 + \lambda)\backslash N]\},$$

where p = state-specific price of intrastate toll service
MC = marginal cost of service
λ = AT&T's conjecture on the effect of its output change on other firms' output
N = industry elasticity of demand (absolute value)
S = AT&T's market share

The estimating equation, following several simplifying assumptions and introducing demand and cost effects, is as follows:

$$\ln P = p_0 + p_1\ \text{ACCESS} + p_2\ \text{REG} + p_3\ \text{SHARE} + p_4 X + p_5 Z + E,$$

where ACCESS = state access charge
REG = the form of regulatory regime (number of months in place)
SHARE = non-AT&T share of the market
X = other variables that affect costs—state-specific CPI, number of employees in the telecommunications sector, per capita
Z = other variables that affect demand—income, CPI, population, and index of industrial performance

Because AT&T's market share is endogenous, it is estimated simultaneously:

24. Robert Kaestner and Brenda Kahn, "The Effects of Regulation and Competition on the Price of AT&T Intrastate Telephone Service," *Journal of Regulatory Economics*, vol. 2 (1990), pp. 363–77.

$$\text{SHARE} = F(\text{ACCESS, REG}, P, \text{EACONV, DENSITY}),$$

where EACONV = percentage of telephones in a state converted to equal access

DENSITY = population density per LATA

RESULTS. The results show the following:

—The number of months a state has implemented either partial flexibility or deregulation significantly lowers AT&T's price; the effect of deregulation is twice as great as the effect of price flexibility. The effects are surprisingly high. Partial flexibility for one year reduced prices 18 percent, all things being equal, and deregulation alone reduces price *42 percent* in one year.

—The effect of deregulation and pricing flexibility operates both through lowering marginal costs (x-inefficiency?) and lowering AT&T's markup.

—A higher market share for AT&T, all things being equal, is associated with a *lower* price. This is partially explained by the equation that endogenizes market share—a 10 percent higher price for AT&T is associated with a 3.3 percent greater market share for competitors. Firms enter where prices are high.

—AT&T's market share falls with deregulation and pricing flexibility. One year of flexibility lowers AT&T's market share by 6 percent; one year of deregulation lowers AT&T's market share by 12 percent.

—A 1 percent change in access costs is associated with a 1.47 percent change in the (day) price. Lowering the access cost by 10 percent therefore lowers prices by 15 percent.

(10) Kahai, Kaserman, and Mayo[25]

Kahai, Kaserman, and Mayo use a two-equation model to test whether competitors are able to constrain AT&T from taking advantage of any residual market power. The two-equation model is as follows:

$$P = P_F(Q_F, P_A, E_A), \tag{1}$$

25. Simran K. Kahai, David L. Kaserman, and John W. Mayo, "Is the 'Dominant Firm' Dominant? An Empirical Analysis of AT&T's Market Power," *Journal of Law and Economics* (forthcoming).

(2) $$P = P_M (Q_M, P_L, T, T^2, Y, D_i),$$

where equation 1 is an inverse supply function relating the market price (taken to be the real price of AT&T's interstate service for a ten-minute, 200-mile call) to (1) Q_F, total interstate switched-access minutes by carriers other than AT&T, and (2) P_A, the real price of access per conversation minute; and E_A, the percentage of lines converted to equal access.

Equation 2, the demand function for *all* minutes (not just of AT&T's competitors), relates the price for interstate calling (as measured by an AT&T 200-mile interstate call to (1) total minutes of interstate calling; (2) the real price of local telephone service (measured by the real CPI); (3) the number of telephones and the number of telephones squared; (4) real disposable per capita income; and (5) D_i, quarterly dummies that pick up seasonal trends in calling. The data are quarterly 1984:3 through 1993:4.

The key result is the coefficient of Q_F—how does the supply of AT&T's competitors relate to AT&T's price? The elasticity estimate at the mean is 4.38. "Thus our results suggest a large supply response to a price change on the part of fringe firms in this industry."[26] Given this estimate, the firm-specific demand elasticity facing AT&T can be estimated as follows:

$$\eta_{\mathrm{ATT}} = \frac{\eta_M}{S_{\mathrm{ATT}}} + \frac{(1 - S_{\mathrm{ATT}})\Sigma_F}{S_{\mathrm{ATT}}},$$

where η_M = industry demand elasticity
η_{ATT} = AT&T's firm-specific demand elasticity
S_{ATT} = AT&T's market share
Σ_F = fringe elasticity of supply

Substituting $\eta_M = .5$ and $S_{\mathrm{ATT}} = .62$ (output basis) or .4 (capacity basis) yields estimates of the firm specific demand elasticity of AT&T of 3.48 to 7.81. The Lerner index ($P - MC/P = 1/\eta_{\mathrm{ATT}}$) then suggests price-cost margins of between .13 and .29—*low* by most standards.

If the supply elasticity is 2.2 (half the estimated elasticity), then AT&T's firm-specific demand elasticity is estimated at between 2.13 and 4.0. The question is whether using quarterly data for nine

26. Ibid.

years does not add some trend to the estimation—a trend not explained by any exogenous supply shift as occurred in their period, when facilities were being rolled out and that may be affecting the estimated coefficients.

(11) MacAvoy[27]

Paul MacAvoy argues that the interstate market is characterized by tacit collusion by the three firms, AT&T, MCI, and Sprint, which accounted for between 86 percent and 97 percent of all interexchange carrier revenues over the 1984–93 period. MacAvoy shows that though AT&T lost market share to MCI and Sprint in the 1984–89 period, the market share of these three competitors stabilized after 1989 (although all three lost market share to resellers). In 1989 rate-of-return regulation of AT&T was replaced by price-cap regulation. MacAvoy suggests that this change in regulatory regime did not accomplish what was intended—to free prices from costs and the inefficiency of the rate-of-return process. Instead, because of the baskets chosen and the rate flexibility allowed, "[AT&T] had sufficient pricing flexibility to discipline MCI and Sprint in response to any strategic discounting undertaken to shift market share in any one category."[28] Combined with asymmetric regulation (such as the advance notice of prices filed by AT&T) and regulatory gaming, an assumption of tacit collusion follows.

The empirical tests of this hypothesis are to develop a set of prices and Herfindahl indexes for MTS as well as for six long-distance markets—switched outbound, dedicated outbound, dedicated outbound–contract, switched inbound, dedicated inbound, and virtual networks. The price data are estimated for representative customers and use tariffed rates.

MacAvoy estimates the price-cost margin by taking incremental costs—the costs of access plus "incremental operating expenses incurred for transporting switched calls from the local exchange to long-distance call receivers,"[29] estimated by WEFA to be $0.01

27. Paul W. MacAvoy, "Tacit Collusion by Regulation in the Pricing of Interstate Long-Distance Services," *Journal of Economics and Management Strategy*, vol. 4 (Summer 1995), pp. 147–85.

28. Ibid., p. 153.

29. Ibid., p. 164.

per minute. We estimate long run incremental costs of long-distance to be between $0.03 and $0.08 per minute (including sales and administrative costs). As Taylor and Taylor 1993 find, the price-cost margin rises over the period.[30] A regression is estimated for these long distance markets relating the weighted average price-cost margin (weights are the providers' market shares) to the HHI and binary service variables as follows (separately for 1987–90 and 1991–93):

$$\Sigma W \frac{(p - \mathrm{MC})}{p} = a + b\,\mathrm{HHI} + \Sigma\, c_i S_i,$$

where $\Sigma W \frac{(p - \mathrm{MC})}{p}$ = weighted price-cost margin

S_i = dummies for each of the seven detailed long-distance markets

MacAvoy finds a negative relationship between the price-cost margin and HHI that is the opposite "[of what] would have been expected if the share-increasing policies of the second and third largest providers drive down pricing margins."[31]

The author's analysis assumes the estimates of price and marginal cost are correct. If prices are overstated and the degree of overestimation increases with time as discounts are applied to all tariffs, and if the long-run marginal costs of use over a ten-year period are above $0.01 per minute and rising because of network changes, then the results are questionable. In addition, if many of the entire costs of interstate telephone service are fixed and sunk, then pricing at a level to recover only operating expenses cannot recover capital costs; hence, pricing above marginal operating costs is necessary. In addition, price discrimination that yields higher margins in less elastic markets is consistent with Ramsey pricing.

30. Taylor and Taylor, "Postdivestiture Long-Distance Competition."
31. MacAvoy, "Tacit Collusion by Regulation," p. 174.

Appendix B

Table 5B-1. Estimates of the Degree of Pass-Through of Access Charges by AT&T in Intrastate Inter-LATA Markets, 1987–93

t Statistics in parentheses

Equation 1 $P_A = \beta_0 + \beta_1\,ACC + \beta_2\,TIME + \beta_3\,DIST + \beta_4\,NOFLEX + \beta_5\,FFLEX + \beta_6\,RORREG + \beta_7\,DENSITY + \beta_8\,RURAL + \beta_9\,BUS + \beta_{10}\,PI/CAP + u$

Equation 2 $P_A = \beta_0 + \beta_2\,TIME + \beta_3\,DIST + \beta_4\,NOFLEX + \beta_5\,FFLEX + \beta_6\,RORREG + \beta_7\,DENSITY + \beta_8\,RURAL + \beta_9\,BUS + \beta_{10}\,PI/CAP + \beta_1^1\,(ACC\text{-}CCL) + \beta_1^2\,CCL(1 - S_A) + u$

Mileage bands	β_0	β_1	β_2	β_3	β_4	β_5	β_6	β_7	β_8	β_9	β_{10}	β_1^1	β_1^2	R^2	*N*
1 and 2	1.50	−0.202	−0.046	0.111	−0.135	0.095	−0.061	−0.081	0.001	−0.017	−0.018			0.36	432
Eq. 1	(16.39)	(−2.95)	(−6.64)	(6.13)	(−3.24)	(4.04)	(−2.83)	(−0.99)	(0.93)	(−3.76)	(−1.87)				
1 and 2	1.47		−0.049	0.111	−0.129	0.094	−0.056	−0.111	0.001	−0.016	−0.015	−0.403	−0.190	0.38	432
Eq. 2	(9.49)		(−7.94)	(6.21)	(−3.12)	(4.07)	(−2.69)	(−1.37)	(1.56)	(−3.74)	(−1.54)	(−4.63)	(0.66)		
3–6	1.36	0.233	−0.049	0.077	−0.071	0.040	−0.049	−0.034	0.002	−0.013	−0.025			0.61	864
Eq. 2	(13.33)	(5.21)	(−10.82)	(14.61)	(−2.60)	(2.59)	(−3.50)	(−0.64)	(3.75)	(−4.70)	(−3.95)				
3–6	1.41		−0.060	0.077	−0.069	0.041	−0.052	−0.060	0.002	−0.012	−0.025	0.043	1.004	0.61	864
Eq. 2	(13.63)		(−14.66)	(14.68)	(−2.54)	(2.69)	(−3.81)	(−1.11)	(4.30)	(−4.27)	(−3.93)	(0.74)	(5.31)		
7 and 8	1.17	0.614	−0.054	0.049	0.028	0.069	−0.051	−0.043	0.003	−0.009	−0.021			0.74	432
Eq. 1	(6.57)	(10.33)	(−9.12)	(3.09)	(0.78)	(3.43)	(−2.73)	(−0.61)	(5.08)	(−2.51)	(−2.52)				
7 and 8	1.22		−0.072	0.049	0.027	0.070	−0.062	−0.087	0.004	−0.008	−0.019	0.412	2.14	0.73	432
Eq. 2	(6.72)		(−12.88)	(3.03)	(0.74)	(3.40)	(−3.34)	(−1.20)	(5.49)	(−2.08)	(−2.23)	(5.29)	(8.36)		
9–11	1.06	0.686	−0.069	0.036	0.061	0.077	−0.042	−0.164	0.006	−0.015	0.002			0.75	648
Eq. 1	(96.79)	(12.20)	(−12.18)	(3.95)	(1.78)	(4.04)	(−2.41)	(−2.45)	(9.46)	(−4.09)	(0.25)				
9–11	1.14		−0.089	0.036	0.057	0.078	−0.056	−0.211	0.006	−0.013	0.003	0.475	2.26	0.73	648
Eq. 2	(7.13)		(−16.75)	(3.86)	(1.62)	(3.99)	(−3.18)	(−3.06)	(9.59)	(−3.54)	(0.33)	(6.42)	(9.27)		
All	1.13	0.349	−0.055	0.074	−0.029	0.064	−0.049	−0.087	0.003	−0.014	−0.014			0.73	2,376
Eq. 1	(16.39)	(11.29)	(−17.54)	(57.01)	(−1.52)	(6.06)	(−5.12)	(−2.35)	(9.02)	(−7.06)	(−3.20)				
All	1.19		−0.068	0.074	−0.028	0.066	−0.056	−0.1185	0.003	−0.012	−0.015	0.147	1.33	0.73	2,376
Eq. 2	(16.95)		(−23.93)	(56.91)	(−1.49)	(6.26)	(−5.86)	(−3.10)	(9.68)	(−6.25)	(−3.30)	(3.68)	(10.18)		

N = Number of observations.

Table 5B-2. Estimates of the Degree of Pass-Through of Access Charges by AT&T in Intrastate Inter-LATA Markets, Allowing for Interactive Effects of Regulation, 1987–93

t Statistics in parentheses

Equation 1 $P_A = \beta_0 + \beta_1$ ACC $+ \beta_2$ TIME $+ \beta_3$ DIST $+ \beta_4$ NOFLEX $+ \beta_5$ FFLEX $+ \beta_6$ RORREG $+ \beta_7$ DENSITY $+ \beta_8$ RURAL $+ \beta_9$ BUS $+ \beta_{10}$ PI/CAP $+ \beta_{11}$ ACC*NOFLEX $+ \beta_{12}$ ACC*FFFLEX $+ \beta_{13}$ ACC*RORREG + u

Equation 2 $P_A = \beta_0 + \beta_2$ TIME $+ \beta_3$ DIST $+ \beta_4$ NOFLEX $+ \beta_5$ FFLEX $+ \beta_6$ RORREG $+ \beta_7$ DENSITY $+ \beta_8$ RURAL $+ \beta_9$ BUS $+ \beta_{10}$ PI/CAP $+ \beta_{11}$ ACC*NOFLEX $+ \beta_{12}$ ACC*FFLEX $+ \beta_{13}$ ACC*RORREG $+ \beta_1^3$ [(ACC-CCL) + CCL(1 − S_A)] + u

Mileage bands	β_0	β_1	β_2	β_3	β_4	β_5	β_6	β_7	β_8	β_9	β_{10}	β_{11}	β_{12}	β_{13}	β_1^3	R^2
1 and 2	1.63	−0.303	−0.048	0.111	−0.462	0.105	−0.173	−0.072	0.000	−0.016	−0.023	0.604	−0.036	0.229		0.73
Eq. 1	(10.01)	(−3.84)	(−6.98)	(6.18)	(−3.45)	(1.52)	(−2.50)	(−0.88)	(0.63)	(−3.78)	(−2.38)	(2.54)	(−0.24)	(1.61)		
1 and 2	1.61		−0.051	0.111	−0.458	0.049	−0.186	−0.111	0.001	−0.016	−0.019	0.804	0.101	0.337	−0.564	0.72
Eq. 2	(10.85)		(−8.16)	(6.28)	(−2.58)	(0.67)	(−2.53)	(−1.38)	(1.30)	(3.69)	(2.00)	(1.85)	(0.47)	(1.61)	(−5.62)	
3–6	1.45	0.190	−0.050	0.077	−0.137	0.031	−0.089	−0.018	0.002	−0.013	−0.029	0.113	0.018	0.080		0.61
Eq. 1	(13.41)	(3.66)	(−11.16)	(14.64)	(−1.52)	(0.68)	(−1.96)	(−0.35)	(3.26)	(−4.54)	(−4.66)	(0.73)	(0.19)	(0.86)		
3–6	1.64		−0.062	0.077	−0.097	0.023	−0.152	−0.057	0.002	−0.012	−0.034	0.006	0.026	0.250	−0.006	0.61
Eq. 2	(16.16)		(−14.83)	(14.50)	(−0.81)	(0.46)	(−3.05)	(−1.04)	(3.26)	(−4.01)	(−5.38)	(0.02)	(0.18)	(1.77)	(−0.09)	
7 and 8	1.24	0.578	−0.053	0.049	0.201	−0.073	−0.017	−0.026	0.003	−0.010	−0.024	−0.328	0.321	−0.104		0.74
Eq. 1	(6.83)	(8.47)	(−8.99)	(3.11)	(1.74)	(−1.23)	(−0.28)	(−0.37)	(4.62)	(−2.55)	(−2.84)	(−1.60)	(2.50)	(−0.84)		
7 and 8	1.56		−0.074	0.049	0.144	−0.018	−0.049	−0.075	0.003	−0.007	−0.034	−0.387	0.251	−0.095	0.472	0.71
Eq. 2	(8.30)		(−12.53)	(2.88)	(0.86)	(−0.26)	(−0.69)	(−0.98)	(3.92)	(−1.84)	(−3.78)	(−0.94)	(1.23)	(−0.48)	(4.95)	
9–11	1.18	0.623	−0.067	0.036	0.379	−0.166	0.009	−0.134	0.005	−0.015	−0.003	−0.602	0.548	−0.164		0.76
Eq. 1	(7.52)	(9.82)	(−12.18)	(4.06)	(3.53)	(−3.00)	(0.15)	(−2.06)	(8.78)	(−4.24)	(−0.34)	(−3.16)	(4.59)	(−1.44)		
9–11	1.48		−0.090	0.036	0.343	−0.053	−0.007	−0.194	0.005	−0.012	−0.012	−0.813	0.384	−0.205	0.545	0.72
Eq. 2	(9.19)		(−16.33)	(3.74)	(2.18)	(−0.82)	(−0.10)	(−2.71)	(7.85)	(−3.28)	(−1.47)	(−2.11)	(2.02)	(−1.10)	(9.19)	
All	1.25	0.289	−0.055	0.074	0.007	−0.028	−0.065	−0.061	0.003	−0.013	−0.020	−0.073	0.208	0.007		0.73
Eq. 1	(17.10)	(8.09)	(−17.72)	(57.28)	(0.12)	(−0.91)	(−2.06)	(−1.66)	(7.83)	(−6.86)	(−4.57)	(−0.68)	(3.09)	(0.12)		
All	1.46		−0.070	0.074	0.001	−0.001	−0.097	−0.107	0.003	−0.012	−0.025	−0.144	0.178	0.079	0.130	0.72
Eq. 2	(21.09)		(−23.95)	(56.12)	(0.01)	(−0.01)	(−2.88)	(−2.84)	(7.70)	(−5.90)	(5.74)	(−0.70)	(1.77)	(0.80)	(2.76)	

Table 5B-3. Estimates of the Degree of Pass-Through of Access Charges by AT&T in Intrastate Inter-LATA Markets, 1987–90

t Statistics in parentheses

Equation 1 $P_A = \beta_0 + \beta_1 ACC + \beta_2 TIME + \beta_3 DIST + \beta_4 NOFLEX + \beta_5 FFLEX + \beta_6 RORREG + \beta_7 DENSITY + \beta_8 RURAL + \beta_9 BUS + \beta_{10} PI/CAP + u$

Equation 2 $P_A = \beta_0 + \beta_2 TIME + \beta_3 DIST + \beta_4 NOFLEX + \beta_5 FFLEX + \beta_6 RORREG + \beta_7 DENSITY + \beta_8 RURAL + \beta_9 BUS + \beta_{10} PI/CAP + \beta_1^1 (ACC\text{-}CCL) + \beta_1^2 CCL(1 - S_A) + u$

Mileage bands	β_0	β_1	β_2	β_3	β_4	β_5	β_6	β_7	β_8	β_9	β_{10}	β_1^1	β_1^2	R^2	N
1 and 2	2.34	−0.412	−0.088	0.126	−0.217	0.110	−0.031	−0.046	−0.001	−0.026	−0.044				
Eq. 1	(9.85)	(−4.39)	(−5.65)	(4.80)	(−4.07)	(2.54)	(−0.74)	(−0.37)	(−0.81)	(−3.48)	(−3.00)			0.39	248
1 and 2	2.13		−0.084	0.126	−0.204	0.110	−0.025	−0.100	−0.000	−0.026	−0.031	−0.582	−0.553		
Eq. 2	(9.03)		(−5.84)	(4.89)	(−3.89)	(2.59)	(−0.60)	(−0.80)	(−0.03)	(−3.60)	(−2.11)	(−5.41)	(−1.28)	0.41	248
3–6	2.08	0.077	−0.089	0.093	−0.145	0.039	−0.008	0.021	0.001	−0.023	−0.048				
Eq. 1	(13.33)	(1.26)	(−8.75)	(12.18)	(−4.21)	(1.39)	(−0.31)	(0.26)	(0.80)	(−4.83)	(−5.07)			0.59	496
3–6	1.96		−0.101	0.093	−0.136	0.040	−0.004	−0.032	0.001	−0.024	−0.038	−0.086	0.968		
Eq. 2	(12.55)		(−10.77)	(12.35)	(−3.97)	(1.45)	(−0.16)	(−0.39)	(1.87)	(−4.96)	(−3.89)	(−1.22)	(3.42)	0.61	496
7 and 8	1.69	0.482	−0.094	0.058	−0.036	0.078	−0.024	0.032	0.003	−0.016	−0.039				
Eq. 1	(6.76)	(6.28)	(−7.41)	(2.70)	(−0.83)	(2.21)	(−0.70)	(0.31)	(3.11)	(−2.57)	(−3.33)			0.72	248
7 and 8	1.61		−0.122	0.058	−0.028	0.080	−0.021	−0.032	0.004	−0.016	−0.029	0.292	2.34		
Eq. 2	(6.48)		(−10.35)	(2.74)	(−0.65)	(2.29)	(−0.62)	(−0.31)	(3.80)	(−2.65)	(−2.40)	(3.31)	(6.57)	0.73	248
9–11	1.58	0.562	−0.114	0.042	−0.002	0.094	−0.015	−0.089	0.005	−0.022	−0.013				
Eq. 1	(7.03)	(7.56)	(−9.28)	(3.30)	(−0.04)	(2.76)	(−0.46)	(−0.89)	(6.20)	(−3.77)	(−1.16)			0.70	372
9–11	1.51		−0.148	0.042	0.004	0.096	−0.013	−0.169	0.006	−0.022	0.328	2.65			
Eq. 2	(6.71)		(−12.82)	(3.32)	(0.10)	(2.82)	(−0.40)	(−1.69)	(7.02)	(−3.84)	(−0.16)	(3.82)	(7.64)		372
All	1.78	0.194	−0.097	0.086	−0.089	0.074	−0.017	−0.019	0.002	−0.022	−0.036				
Eq. 1	(17.34)	(4.72)	(−14.19)	(47.49)	(−4.27)	(3.92)	(−0.94)	(−0.35)	(4.17)	(−6.80)	(−5.70)			0.73	1,364
All	1.67		−0.114	0.086	−0.091	0.075	−0.014	−0.081	0.003	−0.022	−0.025	0.006	1.40		
Eq. 2	(16.21)		(−18.13)	(48.06)	(−3.93)	(4.04)	(−0.75)	(−1.49)	(5.81)	(−6.98)	(−3.87)	(0.12)	(7.35)	0.74	1,364

N = Number of observations.

Table 5B-4. Estimates of the Degree of Pass-Through of Access Charges by AT&T in Intrastate Inter-LATA Markets, 1991–93

t Statistics in parentheses

Equation 1 $P_A = \beta_0 + \beta_1\,ACC + \beta_2\,TIME + \beta_3\,DIST + \beta_4\,NOFLEX + \beta_5\,FFLEX + \beta_6\,RORREG + \beta_7\,DENSITY + \beta_8\,RURAL + \beta_9\,BUS + \beta_{10}\,PI/CAP + u$

Equation 2 $P_A = \beta_0 + \beta_2\,TIME + \beta_3\,DIST + \beta_4\,NOFLEX + \beta_5\,FFLEX + \beta_6\,RORREG + \beta_7\,DENSITY + \beta_8\,RURAL + \beta_9\,BUS + \beta_{10}\,PI/CAP + \beta_1^1\,(ACC\text{-}CCL) + \beta_1^2\,CCL(1 - S_A) + u$

Mileage bands	β_0	β_1	β_2	β_3	β_4	β_5	β_6	β_7	β_8	β_9	β_{10}	β_1^1	β_1^2	R^2	N
1 and 2	0.727	0.024	−0.022	0.091	0.079	0.053	−0.047	−0.152	0.002	−0.015	0.014				
Eq. 1	(3.96)	(0.25)	(−1.39)	(4.52)	(1.11)	(2.25)	(−1.85)	(−1.75)	(2.74)	(−3.21)	(1.29)			0.38	184
1 and 2	0.713		−0.020	0.091	0.058	0.055	−0.047	−0.145	0.002	−0.014	0.013	0.175	−0.148		
Eq. 2	(3.88)		(−1.24)	(4.52)	(0.77)	(2.35)	(−1.85)	(−1.66)	(2.32)	(−3.16)	(1.16)	(0.91)	(−0.45)	0.38	184
3–6	0.727	0.291	−0.014	0.056	0.120	0.013	−0.014	−0.111	0.003	−0.011	0.003				
Eq. 1	(6.36)	(4.93)	(−1.51)	(10.18)	(2.74)	(0.91)	(−0.87)	(−2.07)	(5.66)	(−4.10)	(0.44)			0.56	368
3–6	0.738		−0.016	0.056	0.120	0.013	−0.013	−0.111	0.003	−0.011	0.003	0.286	0.759		
Eq. 2	(6.45)		(−1.67)	(10.16)	(2.61)	(0.88)	(−0.85)	(−2.06)	(5.39)	(−4.08)	(0.43)	(2.41)	(3.74)	0.56	368
7 and 8	0.616	0.661	−0.001	0.036	0.175	0.041	−0.002	−0.163	0.004	−0.013	0.000				
Eq. 1	(2.63)	(6.93)	(−0.08)	(1.78)	(2.48)	(1.75)	(−0.09)	(−1.89)	(4.26)	(−2.96)	(0.65)			0.60	184
7 and 8	0.645		−0.006	0.036	0.184	0.039	−0.002	−0.165	0.004	−0.013	0.007	0.597	1.81		
Eq. 2	(2.74)		(−0.38)	(1.78)	(2.47)	(1.67)	(−0.07)	(−1.90)	(4.17)	(−2.97)	(0.69)	(3.11)	(5.52)	0.60	184
9–11	0.585	0.719	−0.025	0.028	0.231	0.040	−0.005	−0.285	0.006	−0.018	0.028				
Eq. 1	(2.90)	(8.13)	(−1.72)	(2.45)	(3.53)	(1.84)	(−0.21)	(−3.55)	(7.75)	(−4.39)	(2.83)			0.65	276
9–11	0.588		−0.026	0.028	0.209	0.042	−0.004	−0.275	0.006	−0.018	0.027	0.883	1.68		
Eq. 2	(2.91)		(−1.80)	(2.45)	(3.03)	(1.93)	(−0.18)	(−3.42)	(7.05)	(−4.35)	(2.68)	(4.98)	(5.54)	0.66	276
All	0.568	0.427	−0.016	0.057	0.153	0.032	−0.015	−0.175	0.004	−0.014	0.012				
Eq. 1	(6.93)	(9.88)	(−2.31)	(39.76)	(4.78)	(3.10)	(9.96)	(−4.48)	(9.96)	(−6.98)	(2.58)			0.72	1,012
All	0.576		−0.018	0.057	0.144	0.033	−0.015	−0.172	0.004	−0.014	0.012	0.485	1.04		
Eq. 2	(7.02)		(−2.49)	(39.74)	(4.30)	(3.13)	(−1.30)	(−4.37)	(9.23)	(−6.93)	(2.46)	(5.60)	(7.00)	0.72	1,012

N = Number of observations.

Table 5B-5. Estimates of the Degree of Pass-Through of Access Charges by AT&T in Intrastate Inter-LATA Markets, Allowing for Interactive Effects of Regulation, 1987–90

t Statistics in parentheses

Equation 1 $P_A = \beta_0 + \beta_1 \text{ ACC} + \beta_2 \text{ TIME} + \beta_3 \text{ DIST} + \beta_4 \text{ NOFLEX} + \beta_5 \text{ FFLEX} + \beta_6 \text{ RORREG} + \beta_7 \text{ DENSITY} + \beta_8 \text{ RURAL} + \beta_9 \text{ BUS} + \beta_{10} \text{ PI/CAP} + \beta_{11} \text{ ACC*NOFLEX} + \beta_{12} \text{ ACC*FFFLEX} + \beta_{13} \text{ ACC*RORREG} + u$

Equation 2 $P_A = \beta_0 + \beta_2 \text{ TIME} + \beta_3 \text{ DIST} + \beta_4 \text{ NOFLEX} + \beta_5 \text{ FFLEX} + \beta_6 \text{ RORREG} + \beta_7 \text{ DENSITY} + \beta_8 \text{ RURAL} + \beta_9 \text{ BUS} + \beta_{10} \text{ PI/CAP} + \beta_{11} \text{ ACC*NOFLEX} + \beta_{12} \text{ ACC*FFLEX} + \beta_{13} \text{ ACC*RORREG} + \beta_1^S [(\text{ACC-CCL}) + \text{CCL}(1 - S_A)]_t + u$

Mileage bands	β_0	β_1	β_2	β_3	β_4	β_5	β_6	β_7	β_8	β_9	β_{10}	β_{11}	β_{12}	β_{13}	β_1^S	R^2
1 and 2	2.47	−0.548	−0.086	0.126	−0.892	0.284	−0.257	−0.078	−0.001	−0.026	−0.047	−0.320	1.17	0.387		
Eq. 1	(10.38)	(−4.94)	(−5.65)	(4.94)	(−5.02)	(1.64)	(−1.49)	(−0.63)	(−0.84)	(−3.53)	(−3.25)	(−1.02)	(3.98)	(1.28)		0.43
1 and 2	2.19		−0.077	0.126	−0.878	0.126	−0.181	−0.114	−0.000	−0.024	−0.035	−0.043	1.64	0.358	−0.729	
Eq. 2	(10.22)		(−5.34)	(4.96)	(−3.35)	(0.73)	(−1.07)	(−0.93)	(−0.09)	(−3.29)	(−2.48)	(−0.10)	(2.59)	(0.85)	(−5.70)	0.43
3–6	2.13	0.025	−0.088	0.093	−0.491	0.130	−0.097	0.007	0.001	−0.023	−0.049	−0.168	0.604	0.153		
Eq. 1	(13.37)	(0.34)	(−8.71)	(12.27)	(−4.17)	(1.13)	(−0.85)	(0.09)	(0.93)	(−4.88)	(−5.11)	(−0.81)	(3.09)	(0.77)		0.60
3–6	2.23		−0.094	0.093	−0.587	0.126	−0.198	−0.052	0.001	−0.021	−0.051	−0.238	1.03	0.438	−0.181	
Eq. 2	(15.40)		(−9.81)	(12.25)	(−3.36)	(1.10)	(−1.75)	(−0.63)	(0.99)	(−4.37)	(−5.39)	(−0.83)	(2.43)	(1.56)	(−2.13)	0.60
7 and 8	1.72	0.453	−0.094	0.058	−0.033	−0.049	0.037	0.028	0.003	−0.015	−0.041	0.237	−0.011	−0.124		
Eq. 1	(6.74)	(4.86)	(−7.27)	(2.69)	(−0.22)	(−0.34)	(0.26)	(0.27)	(2.91)	(−2.44)	(−3.40)	(0.90)	(−0.04)	(−0.49)		0.72
7 and 8	2.09		−0.117	0.058	−0.252	0.055	−0.057	−0.055	0.002	−0.014	−0.051	0.040	0.409	0.018	0.279	
Eq. 2	(8.20)		(−9.09)	(2.55)	(−1.07)	(0.36)	(−0.37)	(−0.50)	(2.40)	(−2.16)	(−4.02)	(0.10)	(0.72)	(0.05)	(2.44)	0.69
9–11	1.64	0.505	−0.113	0.042	0.194	−0.265	0.153	−0.091	0.005	−0.020	−0.017	0.669	−0.360	−0.339		
Eq. 1	(7.26)	(5.69)	(−9.19)	(3.35)	(1.36)	(−1.91)	(1.11)	(−0.93)	(5.73)	(−3.46)	(−1.51)	(2.67)	(−1.52)	(−1.40)		0.72
9–11	2.04		−0.144	0.042	0.085	−0.008	0.026	−0.191	0.005	−0.020	−0.026	0.238	−0.359	−0.177	0.338	
Eq. 2	(9.01)		(−11.61)	(3.11)	(0.37)	(−0.05)	(0.18)	(−1.80)	(5.21)	(−3.24)	(−2.13)	(0.64)	(−0.66)	(−0.49)	(3.06)	0.67
All	1.85	0.130	−0.096	0.086	−0.294	0.018	−0.033	−0.031	0.002	−0.021	−0.038	0.107	0.333	0.011		
Eq. 1	(17.46)	(2.61)	(−13.94)	(47.58)	(−3.69)	(0.23)	(−0.43)	(−0.57)	(3.95)	(−6.58)	(−5.97)	(0.76)	(2.52)	(0.08)		0.74
All	2.01		−0.109	0.086	−0.396	0.076	−0.108	−0.101	0.002	−0.020	−0.041	−0.022	0.647	0.179	−0.056	
Eq. 2	(20.87)		(−16.68)	(47.19)	(−3.33)	(0.98)	(−1.40)	(−1.82)	(4.06)	(−6.12)	(−6.42)	(−0.11)	(2.26)	(0.94)	(−0.96)	0.73

Table 5B-6. Estimates of the Degree of Pass-Through of Access Charges by AT&T in Intrastate Inter-LATA Markets, Allowing for Interactive Effects of Regulation, 1991–93

t Statistics in parentheses

Equation 1 $P_A = \beta_0 + \beta_1$ ACC $+ \beta_2$ TIME $+ \beta_3$ DIST $+ \beta_4$ NOFLEX $+ \beta_5$ FFLEX $+ \beta_6$ RORREG $+ \beta_7$ DENSITY $+ \beta_8$ RURAL $+ \beta_9$ BUS $+ \beta_{10}$ PI/CAP $+ \beta_{11}$ ACC*NOFLEX $+ \beta_{12}$ ACC*FFFLEX $+ \beta_{13}$ ACC*RORREG + u

Equation 2 $P_A = \beta_0 + \beta_2$ TIME $+ \beta_3$ DIST $+ \beta_4$ NOFLEX $+ \beta_5$ FFLEX $+ \beta_6$ RORREG $+ \beta_7$ DENSITY $+ \beta_8$ RURAL $+ \beta_9$ BUS $+ \beta_{10}$ PI/CAP $+ \beta_{11}$ ACC*NOFLEX $+ \beta_{12}$ ACC*FFLEX $+ \beta_{13}$ ACC*RORREG $+ \beta_1^3$ [(ACC-CCL) + CCL(1 − S_A)] + u

Mileage bands	β_0	β_1	β_2	β_3	β_4	β_5	β_6	β_7	β_8	β_9	β_{10}	β_{11}	β_{12}	β_{13}	β_1^3	R^2
1 and 2	0.964	−0.251	−0.025	0.091	0.024	−0.127	−0.260	−0.131	0.002	−0.016	0.010	0.480	0.076	0.506		
Eq. 1	(4.93)	(−1.99)	(−1.61)	(4.62)	(0.10)	(−1.51)	(−3.19)	(−1.52)	(2.26)	(−3.55)	(0.90)	(2.22)	(0.13)	(2.66)		0.38
1 and 2	0.894		−0.023	0.091	0.062	−0.145	−0.241	−0.146	0.002	−0.017	0.014	0.682	−0.072	0.612	−0.259	
Eq. 2	(4.61)		(−1.51)	(4.59)	(0.31)	(−1.54)	(−2.50)	(−1.69)	(2.17)	(−3.69)	(1.34)	(2.15)	(−0.13)	(2.00)	(−1.34)	0.41
3–6	0.885	0.129	−0.017	0.056	0.515	−0.143	−0.122	−0.089	0.003	−0.013	0.000	0.417	−1.00	0.265		
Eq. 1	(7.32)	(1.68)	(−1.77)	(10.46)	(3.45)	(−2.78)	(−2.44)	(−1.69)	(5.18)	(−4.68)	(0.02)	(3.15)	(−2.84)	(2.28)		0.59
3–6	0.846		−0.026	0.056	0.478	−0.147	−0.121	−0.098	0.002	−0.013	0.002	0.560	−1.12	0.356	0.259	
Eq. 2	(7.09)		(−1.66)	(10.42)	(3.91)	(−2.56)	(−2.06)	(−1.85)	(4.61)	(−4.74)	(0.27)	(2.89)	(−3.39)	(1.91)	(2.18)	0.59
7 and 8	0.851	0.469	−0.005	0.036	0.700	−0.248	−0.066	−0.117	0.003	−0.016	0.002	0.767	−1.30	0.148		
Eq. 1	(3.62)	(3.84)	(−0.35)	(1.87)	(2.97)	(−3.04)	(−0.84)	(−1.40)	(3.75)	(−3.59)	(0.22)	(3.67)	(−2.34)	(0.80)		0.65
7 and 8	0.803		−0.004	0.036	0.668	−0.225	−0.064	−0.133	0.003	−0.015	0.003	0.937	−1.53	0.202	0.775	
Eq. 2	(3.36)		(−0.29)	(1.82)	(3.38)	(−2.42)	(−0.67)	(−1.56)	(3.01)	(−3.46)	(0.31)	(2.99)	(−2.86)	(0.67)	(4.05)	0.63
9–11	0.791	0.585	−0.029	0.028	0.780	−0.253	0.003	−0.234	0.006	−0.020	0.023	0.776	−1.34	−0.028		
Eq. 1	(3.83)	(5.12)	(−2.06)	(2.55)	(3.54)	(−3.31)	(0.04)	(−3.00)	(7.20)	(−5.05)	(2.47)	(3.96)	(−2.57)	(−0.16)		0.68
9–11	0.735		−0.026	0.028	0.745	−0.247	0.007	−0.241	0.005	−0.021	0.025	1.02	−1.56	−0.039	1.001	
Eq. 2	(3.61)		(−1.88)	(2.56)	(4.13)	(−2.92)	(0.08)	(−3.11)	(6.14)	(−5.05)	(2.51)	(3.58)	(−3.21)	(−0.14)	(5.74)	0.69
All	0.768	0.246	−0.019	0.057	0.532	−0.189	−0.103	−0.141	0.003	−0.016	0.009	0.590	−0.952	0.207		
Eq. 1	(8.82)	(4.36)	(−2.81)	(40.83)	(4.87)	(−5.01)	(−2.82)	(−3.66)	(8.93)	(−7.96)	(1.82)	(6.09)	(−3.69)	(2.44)		0.74
All	0.717		−0.018	0.057	0.510	−0.188	−0.097	−0.152	0.003	−0.016	0.010	0.777	−1.12	0.267	0.461	
Eq. 2	(8.35)		(−2.59)	(40.73)	(5.70)	(−4.48)	(−2.27)	(−3.95)	(7.76)	(−8.03)	(2.18)	(5.48)	(−4.65)	(1.96)	(5.32)	0.73

Chapter Six

Regulatory Reform and the Intrastate-Intraprovincial Rate Structure

WE NOW TURN TO AN analysis of the market for services provided by the local-exchange companies in the United States and by the intraprovincial operations of the Canadian telephone companies, focusing particularly on the effects of regulatory reform and competition on intrastate, intra-LATA (intraprovincial) toll charges and local subscriber rates. If the purpose of reform and market liberalization is to move rates closer to incremental costs, or even to Ramsey prices, we should expect to see residential access rates rise relative to business rates, access rates in less dense areas rise relative to urban access rates, and intra-LATA long-distance rates fall sharply. To see whether these changes have occurred, we examine the changes in rates between 1980 (or 1983) and 1993 and the dispersion in rates across U.S. states (and, to a lesser extent, Canadian provinces) during 1987–93, the period in which regulatory institutions were changing rapidly in the United States and beginning to change in Canada. These results complement our analysis of the effects of competition and regulation in the United States inter-LATA markets reported in chapter 5.

Local Access Rates

Rates for customer access to the telephone network, "local" rates, are regulated in the United States by state regulatory commissions, subject to a variety of federal requirements imposed by the Federal Communications Commission. The monthly rate for connecting to the network is the sum of state-regulated charges and the federal subscriber line charge. In most U.S. jurisdictions,

Table 6-1. Average Residential and Business Flat Monthly Rates, December 1980 versus December 1993[a]

$ per month

	Residential		*Multiline business*		*Ratio business to residential*	
Area	*1980*	*1993*	*1980*	*1993*	*1980*	*1993*
Rural	6.58(44)	14.75(42)	14.25(43)	31.88(40)	2.18(42)	2.22(40)
Urban	9.78(42)	17.63(42)	32.55(43)	41.68(37)	2.65(35)	2.56(37)
Ratio urban to rural	1.49(42)	1.20(42)	1.83(36)	1.30(37)		

Sources: National Association of Regulatory Utility Commissioners, *Bell Operating Companies' Exchange Rates*; FCC.

a. Numbers in parentheses indicate number of states with published rates used to construct the average or the ratio.

residences and businesses have an option of various flat-rate and measured-service plans.[1] In Canada, local residential and business access is offered only on a flat rate, and the service is typically offered over very large local calling areas, especially in the Bell Canada territory.[2]

The United States

For the current, twisted-pair local-loop technology, the cost of providing subscriber access is generally much higher in rural areas than in dense urban areas. As we showed in chapter 3, despite the cost differential, urban rates are generally higher than rural rates. Moreover, the recent trend toward some state regulatory reform has not had much of an effect on the rural-urban rate discrimination at the state level in the United States. Rate rebalancing has not yet begun in Canada.

Table 6-1 provides summary information on monthly flat rates for local service in the largest exchanges and in the smallest ex-

1. Flat rates are those that allow unlimited local calling within a designated local calling area. Measured service generally requires a lower monthly flat rate plus a charge for each minute of calls originated by the customer in excess of a stated minimum number of monthly minutes. Some states have a number of measured-service options; very few rely solely on flat rates.

2. One should remember that the CRTC regulated only Bell Canada and British Columbia Telephone until 1992. Therefore, local rates were controlled by provincial regulators in the other provinces. In addition, there is no subscriber line charge anywhere in Canada.

changes for the U.S. Bell operating companies in 1980 and 1993. The urban and rural rates represent the applicable rates in the largest and the smallest exchanges, respectively, in each state. The largest exchanges are obviously to be found in the center of the most densely populated areas, while the smallest exchanges are likely to exist only in areas of low density because of the economies of scale that are realized in switching systems. Rural rates have tended to rise somewhat more rapidly than have urban rates in the United States since 1980, but the ratio of urban rates to rural rates remains substantially above unity despite the fact that the ratio of the incremental cost of service is more likely to be in the range of 0.2 to 0.3.[3]

Nor is there any evidence that states that are experimenting with new, less intrusive forms of regulation are any different with regard to urban-rural rate differentials than the more rigid states. By 1993, twenty-seven states had authorized some form of rate cap, deregulation, or revenue-sharing regulation, but these states exhibited no statistically significant difference from the others in the rate of change in urban-rural residential rates between 1980 and 1993 (table 6-2). The most notable declines in urban-rural discrimination are to be found in Colorado, Nevada, and Nebraska. Colorado and Nevada have adopted quite different revenue-sharing plans, and Nebraska is the only fully deregulated state.[4]

Most studies show that the fixed cost of connecting business subscribers to the telephone network is not significantly greater than the fixed cost of residential connections.[5] Businesses use the network more intensively, but the additional fixed costs of use are very low relative to the cost of the loop itself. Thus business rates should be marginally greater than residential rates, but not two to three times as great. Data for the largest and smallest exchanges in each state (table 6-1) show that U.S. businesses pay more than twice the rate charged residences for flat-rate service. Moreover, this differential appears to be rising somewhat in the rural areas and falling only gradually in the more urbanized areas. In 1980, in

3. See chapter 3, table 3-1.

4. Nebraska is deregulated in the sense that direct rate supervision has been lifted. However, entry is still barred into local wire-line service.

5. See, for example, Bridger M. Mitchell, *Incremental Costs of Telephone Access and Local Use* (RAND Corporation, July 1990).

Table 6-2. Ratio of Urban to Rural Monthly Flat Rates in States with Regulatory Flexibility, 1980 and 1993[a]

	Residential		*Business*	
State	*1980*	*1993*	*1980*	*1993*
Alabama	1.44	1.25	1.69	1.41
California	1.59	1.00	1.69	n.a.
Colorado	1.40	1.00	2.15	1.01
Connecticut	1.35	1.29	1.35	1.33
Florida	1.43	1.31	1.46	1.36
Georgia	1.71	1.32	2.42	1.65
Idaho	1.37	1.14	1.71	1.17
Kentucky	1.49	1.37	1.72	1.58
Louisiana	1.49	1.26	1.61	1.36
Maryland	1.52	1.09	n.a.	n.a.
Michigan	1.31	1.25	1.31	1.00
Minnesota	1.25	1.10	1.58	1.31
Mississippi	1.44	1.22	1.67	1.34
Missouri	1.76	1.45	2.76	1.91
Nebraska	1.61	1.05	1.81	1.03
Nevada	1.38	1.00	1.73	1.00
New Jersey	1.23	1.14	1.11	n.a.
New York	2.02	1.62	2.13	n.a.
New Mexico	1.36	1.35	1.51	1.44
North Dakota	1.35	1.17	1.35	1.18
Oregon	1.18	1.00	1.78	1.00
Rhode Island	1.36	1.31	1.27	n.a.
South Carolina	1.47	1.15	1.69	1.25
Tennessee	1.66	1.42	2.05	1.39
Texas	1.66	1.25	2.25	1.36
Washington	1.60	1.16	2.38	1.44
Wisconsin	1.06	n.a.	1.33	n.a.
Average of above	1.57(26)	1.24(26)	1.88(21)	1.34(21)
Average for all states	1.56(42)	1.22(42)	1.87(37)	1.34(37)

Source: National Association of Regulatory Utility Commissioners, Washington, December 1993.

a. States with revenue sharing, rate caps, or deregulation (Wisconsin has a hybrid system). Numbers in parentheses indicate number of states with published rates in both years used to construct weighted average, with number of access lines as weights.

n.a. Not applicable (no flat rate for one or more services.)

the largest exchanges (urban areas), flat business rates were an average of 2.65 times higher than residential flat rates. By 1993 they were still an average of 2.56 times greater than residential rates. In 1980 business rates were 2.18 times higher than residential rates in the rural areas, but by 1993 they had actually risen to 2.22 times residential rates in these smaller exchanges.[6]

6. Note that the comparisons in table 6-1 are for those states for which the National Association of Regulatory Utility Commissioners (NARUC) reports flat rates.

Table 6-3. Ratio of Business to Residential Monthly Flat Rates in States with Regulatory Flexibility, 1980 and 1993

State	*Urban* 1980	*Urban* 1993	*Rural* 1980	*Rural* 1993
Alabama	2.95	2.65	2.51	2.35
California	2.72	n.a.	2.55	n.a.
Colorado	2.97	2.30	1.93	2.28
Connecticut	3.00	2.55	3.00	2.44
Florida	2.42	2.48	2.37	2.39
Georgia	2.79	2.44	1.97	1.95
Idaho	2.43	2.20	1.94	2.13
Kentucky	2.89	2.68	2.51	2.32
Louisiana	2.89	2.55	2.67	2.37
Maryland	n.a.	n.a.	n.a.	n.a.
Michigan	1.95	0.98	1.95	1.22
Minnesota	3.08	2.60	2.44	2.16
Mississippi	2.54	2.37	2.20	2.16
Missouri	2.97	2.63	1.89	1.99
Nebraska	2.31	2.28	2.05	2.31
Nevada	2.66	2.58	2.12	2.59
New Jersey	1.95	n.a.	2.16	1.75
New York	1.84	n.a.	1.75	1.91
New Mexico	1.72	2.96	1.55	2.77
North Dakota	2.00	2.25	2.00	2.20
Oregon	2.75	2.24	1.82	2.24
Rhode Island	2.35	n.a.	2.53	2.35
South Carolina	2.81	2.46	2.44	2.26
Tennessee	2.77	2.90	2.24	2.96
Texas	3.37	2.34	2.49	2.15
Washington	2.91	2.24	1.96	1.81
Wisconsin	2.64	n.a.	2.11	n.a.
Average of above	2.78(21)	2.32(21)	2.26(21)	2.15(21)
Average for all states	2.71(37)	2.37(37)	2.23(37)	2.18(37)

Source: NARUC.

a. States with revenue sharing, rate caps, or deregulation. Numbers in parentheses indicate number of states with published rates in both years used to construct weighted average, with number of access lines as weights.

n.a. Flat rate not available for one or more services.

Nor does regulatory flexibility appear to have a significant effect on the differential between business and residential rates (table 6-3). In fact in the only deregulated state, Nebraska, the differential has risen in rural areas, perhaps reflecting the carrier's belief that demand for business access is less price elastic than residential demand.

A final indication of reform in the pricing of local telephone service in the United States is the offering of measured service to residential consumers even if the findings of Rolla Edward Park and Bridger Mitchell question whether such an offering measurably improves welfare.[7] In 1980 only twenty states allowed local exchange companies to offer some type of measured residential service.[8] By 1993 every state except Maine (of the lower forty-eight states) had begun to offer residential subscribers some form of measured service. Given the divergence of state approaches to regulation and the nearly ubiquitous nature of measured service options, this movement to more choices among rate plans is obviously not solely the result of deregulation or liberalization; it simply reflects other forces at work on the state public utility commission.

Canada

Until recently, telephone regulators in Canada have been much less inclined to permit new entry than their U.S. counterparts. Through 1992, competition existed only in terminal equipment, private lines, and the resale and shared use of monopoly long-distance switched-message traffic. Entry by facilities-based carriers into switched long-distance services has now been approved by the CRTC, but the full effects of this entry have not yet been registered. Private firms cannot erect their own microwave facilities to bypass the monopoly network. Entry by new local access-exchange carriers has not yet occurred, although the CRTC opened the doors wide to such competition in 1994, inviting new entrants and requiring substantial unbundling of local telephone networks.[9]

Canadian regulatory officials, only recently dealing with liberalization of local service, are just beginning the transition to new regulatory regimes to restrain costs and prevent cross subsidization. In Decision 94-19, the CRTC announced its intention to shift to a rate-cap regime in three years after the imposition of a substantial rebalancing of rates. Specifically, the CRTC ordered that $2 be added to monthly local flat rates in each of the next three years and that long-distance rates be reduced accordingly.[10]

7. Rolla E. Park and Bridger M. Mitchell, "Optimal Peak-Load Pricing for Local Telephone Calls," RAND Corporation, March 1987.

8. These measured-service plans vary substantially across the states and are referred to as "message rate" service or "local measured service" in most jurisdictions.

9. CRTC Decision #94-19.

10. This total of $6 per line per month is to be added to local rates to allow the

Unfortunately, the CRTC's bold new repricing strategy was placed on hold by the Canadian cabinet because it was opposed by the new long-distance carriers, such as Unitel, who were to be subsidized by being offered access rates for originating and terminating calls that were lower than the rates that Bell Canada and the Stentor partners would be forced to impute on their own calls. A high contribution rate assigned to long-distance calls increases the magnitude of this subsidy to the struggling new carriers. The entire issue was referred back to the CRTC, which subsequently ruled to impose $4 of the original $6 in subscriber line charges.

The once staid regulatory climate in Canada provides a useful control sample for any analysis of the effects of the regulatory changes in the United States. Canada shares with the United States a long border, similar technology, and a similar standard of living. Its telecommunications regulatory policies and the preponderance of remote service areas are arguably the only major differences in the telephone environment between the two countries. Given Canada's late start in liberalizing its telephone sector, it should be interesting to see if its telephone rates are significantly different from those in the United States. Table 6-4 provides a clear answer for 1990 local rates: Canadian residential rates and rural business rates were dramatically lower than their U.S. counterparts. Even if the CRTC eventually succeeds in imposing a C$6 per month surcharge on all lines, a substantial differential will remain in many provinces. Given that U.S. rural rates are far too low relative to rates in larger switching centers, the data in table 6-4 suggest that Canada has far more repricing to accomplish if it is eventually to open up all markets to competition.

The business-residential discrimination in Canada has been far more severe than in the United States. It is thus not surprising that much of the momentum for liberalization and deregulation of Canadian telecommunications has come from the business sector. The low residential flat rates can only be supported from very high

---

access charge or contribution from long-distance services to be reduced so that the new competitive long-distance carriers do not seek to bypass the local carriers' networks. A similar policy was launched in the United States by the FCC in 1984, but the shift of costs was subsequently limited to a maximum of $3.50 per month for residential subscribers and single-line businesses. Nevertheless, as we shall see, the proposed $6 per month shift will not result in local Canadian rates' rising to U.S. levels.

Table 6-4. Canadian Local Flat Monthly Rates, 1994
US$ per month[a]

	Residential		*Business*	
Province	*Urban*	*Rural*	*Urban*	*Rural*
Alberta	10.24	8.77	26.34	16.42
British Columbia	15.96	5.23	53.30	9.99
Manitoba	9.08	5.67	23.32	10.58
New Brunswick	9.70	8.38	25.40	15.96
Newfoundland	9.85	6.63	32.21	16.11
Nova Scotia	12.37	10.87	39.10	20.68
Ontario-Quebec	13.91	4.10	35.14	11.71
Prince Edward Island	11.86	10.14	35.95	23.50
Saskatchewan	11.24	9.26	24.56	16.95
Canadian average	11.58	7.67	32.81	15.77
U.S. average (December 1993)[b]	13.86	11.60	33.35	25.64

Sources: Provincial telephone companies' tariffs; NARUC.
a. All rates converted at the 1994 exchange rate.
b. Urban rates for forty-two states; rural data for thirty-six states. See table 6-16.

long-distance rates. Business accounts for a large share of long-distance calling, so the pressure from business for reform is all the more understandable.

Intra-LATA and Intraprovincial Long-Distance Rates

As a result of the AT&T antitrust settlement, the divested Bell operating companies are confined to long-distance services within their LATAs and the interexchange carriers are the only suppliers of intrastate services between LATAs. This settlement led to the development of competition in the inter-LATA markets within the thirty-eight multi-LATA states despite the hostility of traditional state regulatory commissions to such competition. For the intra-LATA market, however, the pace of competition has been much slower. The local telephone companies, including the BOCs, dominate intra-LATA service because they automatically receive all 1+ MTS dialed service in their LATAs and generally are not required to hand it off to other carriers.[11] In Canada, there has been no such structural separation; therefore, the Stentor companies continue to dominate intraprovincial calling.

11. This situation began to change in 1993.

Table 6-5. Average U.S. BOC Intra-LATA Daytime Rates, December[a]
$ per five-minute call

				Percent change	
Mileage	*1980*	*1983*	*1993*	*(1980–83)*	*(1983–93)*
0–10	0.48	0.56	0.57(47)	16.7	1.8
11–16	0.65	0.76	0.74(47)	16.9	−2.6
17–22	0.81	0.93	0.88(47)	14.8	−5.4
23–30	0.99	1.15	1.05(47)	16.2	−8.7
31–40	1.19	1.35	1.15(47)	13.4	−14.8
41–55	1.36(47)	1.54(47)	1.26(47)	13.2	−18.2
56–70	1.51(47)	1.71(47)	1.36(47)	13.2	−20.5
71–124	1.70(47)	1.88(47)	1.43(47)	10.6	−23.9
125–196	1.89(39)	2.06(40)	1.56(40)	9.0	−24.3
197–292	2.07(38)	2.19(36)	1.62(36)	5.8	−26.0

Source: NARUC.
a. Numbers in parentheses indicate number of states in cell. All others are for forty-eight mainland states.

U.S. RATES. The changes in the pattern of U.S. daytime intra-LATA rates are shown in table 6-5. Between 1980 and 1983, the Bell operating companies' intra-LATA rates rose, but since 1983 the rates for a five-minute call in all but the shortest mileage band have declined substantially. Note the sharp reduction in the distance premium over both periods, but the distance premium is still much greater than that found in the more competitive inter-LATA markets (see table 5-3).

The development of greater competition in intrastate markets has generated an interest in more flexible regulation of intra-LATA and inter-LATA rates. Although the states have been much slower to admit greater rate flexibility in long-distance rates than has the FCC, there has been a substantial movement away from formal proceedings for every adjustment in intrastate rates. It is unclear, however, whether this flexibility has led to lower rates or to greater carrier efficiency.

In some states, reselling of the local-exchange carrier's intra-LATA services is permitted, but no facilities-based carriers, such as AT&T or MCI, may offer service. In others, both facilities-based carriers and resellers are permitted. Once competition is admitted, the regulators face the problem of deciding on the degree of rate freedom to allow regulated carriers, particularly the dominant local-exchange carrier (LEC) or AT&T. Basically, there are three choices for the regulatory commission: allow full price flexibility (deregulation), allow flexibility within a given range (banded regulation), or require a full investigation of each tariff change (full regulation). Those states that decide not to admit competition into intra-LATA service instead allow limited competition by permitting resellers or arbitrageurs to buy bulk LEC offerings and resell them at rates of their own choosing.

How have changes in regulatory regime affected intra-LATA rates? Table 6-6 shows the average rates for five-minute daytime calls in each mileage band in December 1983 and December 1993 for states electing each of three regulatory options: facilities-based competition with rate flexibility; facilities-based competition with no rate flexibility; and reselling but no facilities-based competition. The pattern is quite striking.

First, those states that have allowed competition and permitted local carriers to have flexible rates have experienced smaller rate reductions than those that have not permitted rate flexibility, and the differences are statistically significant in most mileage bands. Second, states that allow competition without pricing flexibility had roughly the same rates in 1983 as those now permitting rate flexibility. Finally, those states still prohibiting competition (only six in 1993) had much lower rates in 1983 than those that have since liberalized; thus their current rates are generally no higher than those that have allowed competition. The degree of rate progression as the distance of the call increases is greatest for the noncompetitive states.

The above results are at apparent odds with the general results of the FCC and FTC studies of the effects of liberalization.[12] These studies pool data for all states in all mileage bands, even though there are as many as forty-eight observations in most bands, obtaining a statistically significant relationship between rates and a much narrower choice of regulatory variables. Our result in table 6-6 suggests that rates are not generally lower in states with intra-

12. Alan D. Mathios and Robert P. Rogers, "The Impact of State Price and Entry Regulation on Intrastate Long-Distance Telephone Rates," Federal Trade Commission, November 1988; Alan D. Mathios and Robert P. Rogers, "The Impact of Alternative Forms of State Regulation of AT&T on Direct-Dial Long-Distance Telephone Rates," *RAND Journal of Economics*, vol. 20, no. 3 (Autumn 1989); Chris Frentrup, "The Effect of Competition and Regulation on AT&T's Intrastate Toll Prices and of Competition on Bell Operating Company Intra-LATA Toll Prices," Industry Analysis Division, Common Carrier Bureau, Federal Communications Commission, Washington, June 1988.

Table 6-6. Average Daytime Intra-LATA Long-Distance Rates for BOCs by Type of State Regulatory Regime, 1993
$ per five-minute call

	States allowing competition					
	With pricing flexibility		*Without pricing flexibility*		*States prohibiting competition*	
Mileage	*1983*	*1993*	*1983*	*1993*	*1983*	*1993*
0–10	0.55	0.65	0.59[a]	0.55[a]	0.44	0.49
11–16	0.72	0.82	0.80[a]	0.72[a]	0.63	0.69
17–22	0.90	0.99	0.96[a]	0.86[a]	0.79	0.76
23–30	1.18	1.23	1.16[a]	0.99[a]	0.99	0.99
31–40	1.39	1.35	1.37[a]	1.09[a]	1.15	1.07
41–55	1.55	1.43	1.58[b]	1.21[b]	1.30	1.15
56–70	1.72	1.53	1.76[b]	1.29[b]	1.47	1.34
71–124	1.86	1.59	1.95[b]	1.36[b]	1.60	1.44
Number of states	12	12	30[a,b]	28[a,b]	6	6

Sources: NARUC; *State Telephone Regulation Report.*
a. Rate for Louisiana not available in 1993. Average for twenty-nine states.
b. Excludes Rhode Island. Average for twenty-nine states.

LATA competition and that pricing flexibility is associated with somewhat higher rates than those in states with no flexibility.

On reflection, the data in table 6-6 may not be altogether surprising. Long-distance carriers may be able to penetrate the large business intra-LATA market through WATS or other bulk offerings, but they must find it both less attractive and more difficult to compete for residential and smaller businesses' traffic, given the LEC's grasp on all directly dialed traffic. As of late 1994, only six states required some local-exchange carriers to provide full parity in dialing intra-LATA service to customers of competitive interexchange carriers.[13] As a result, the IX (long-distance) carriers competing with the local company for intra-LATA MTS traffic could not easily gain access to small users, nor are they likely to find such users a very attractive market.[14] Since the data in table 6-6 are based on the regularly tariffed MTS charges, they reflect the degree of competition in the market for these smaller users. Competition's major effect may be in the market for large business users, al-

13. *State Telephone Regulation Report* (Alexandria, Va.: Telecom Publishing Group, October 20 and November 3, 1994).

14. In our analysis of the inter-LATA market in chapter 5, by contrast, discount plans for the larger toll users were a more important phenomenon.

Table 6-6. (contd.)

States allowing competition in 1993			*States prohibiting competition in 1993*
With pricing flexibility	*Without pricing flexibility*		
California	Alabama	Mississippi	Arizona
Delaware	Arkansas	Missouri	Nevada
Idaho	Colorado	Montana	New Jersey
Kansas	Connecticut	New Hampshire	Oklahoma
Michigan	Florida	New York	Virginia
Nebraska	Georgia	North Carolina	Wyoming
New Mexico	Illinois	Ohio	
North Dakota	Indiana	Oregon	
South Carolina	Iowa	Pennsylvania	
South Dakota	Kentucky	Rhode Island	
Vermont	Louisiana	Tennessee	
West Virginia	Maine	Texas	
	Maryland	Utah	
	Massachusetts	Washington	
	Minnesota	Wisconsin	

though discount plans were much less important in the intra-LATA market than in inter-LATA markets in the 1987–91 period.[15]

The apparently perverse effects of pricing flexibility may not be so inexplicable if the above deductions about LEC dominance of intra-LATA traffic are accurate. If liberalized entry attracts few new carriers in smaller LATAs with mostly short-distance traffic, promoting competitive entry and full price flexibility may have a perverse effect since the dominant carriers remain dominant but are deregulated.[16] Moreover, if state regulatory commissions allow the LECs to discriminate in favor of their own intra-LATA service or to set intra-LATA access charges sufficiently high, competition from non-LEC carriers may have little effect on rates. Indeed, rates rose most dramatically in the shorter mileage bands in the states with full price flexibility and competition.

Note that the above inferences are generally applicable to the 1993 evening and night/weekend rates for ordinary consumer MTS

15. National Association of Regulatory Commissioners, *Bell Operating Companies' Long-Distance Message Telephone Rates*, annual editions.

16. Note that these results are different from those obtained for the *inter-LATA* market in chapter 5, where we found that full flexibility is effective in reducing rates. This difference may reflect the differences in access arrangements across the two markets. Equal access is now virtually ubiquitous in inter-LATA markets; as of 1993, equal access was not available in any state's intra-LATA market.

service. Table 6-7 shows that for these periods the longer-distance intra-LATA rates are lower in states allowing competition but less than full pricing flexibility than in those allowing full flexibility. States prohibiting competition have generally lower evening rates and night/weekend rates than do competitive states. Thus once again there is some evidence of a perverse effect of full rate flexibility and even of competition.

CANADIAN LONG-DISTANCE RATES. Long-distance service within Canada may be divided into intraprovincial and interprovincial service. The provinces had regulated the former—with the exception of Ontario, Quebec, and British Columbia—until the recent passage of the 1990 Telecommunications Act. The CRTC has traditionally regulated interprovincial rates, although our discussion in chapter 2 may suggest that such regulation was at best informal.

This chapter analyzes interprovincial and intraprovincial long-distance services because both are still offered by—indeed, dominated by—local-exchange carriers in the Stentor group. Interprovincial daytime rates for 1980 and 1994 are displayed in table 6-8.[17] These average rates are for all member companies in 1980 and a subset of them in 1994.[18] Note that rates for all mileage bands have fallen substantially in nominal dollars since 1980, but that they remain far above AT&T's 1994 U.S. consumer interstate rates. For all mileage bands greater than 290 miles, Canadian rates are at least 20 percent higher than AT&T's rates for early 1994. Even in the 57–180 mileage bands, Canadian rates are 8 to 20 percent higher than U.S. rates.

Surprisingly, intraprovincial rates are closer to U.S. levels. As table 6-9 shows, intraprovincial daytime MTS rates in only one Canadian province—Nova Scotia—were above the U.S. average in 1994, and then only for the longer mileage bands. Three others—Prince Edward Island, Newfoundland, and New Brunswick—had generally similar rates. The remaining five provinces[19] had generally lower rates, except in the very shortest mileage bands.

17. Calls between Ontario and Quebec are deemed intraprovincial calls because Bell Canada's service area spans the two provinces.

18. See the note to table 6-8.

19. Ontario and Quebec rates are combined because Bell Canada reports them together.

Table 6-7. Average Evening and Night/Weekend Intra-LATA Long-Distance Rates for BOCs by Type of State Regulatory Regime, 1993[a]

U.S.$ per five-minute call

	States allowing competition				*States prohibiting competition*	
	With pricing flexibility		*Without pricing flexibility*			
Mileage	*Evening*	*Night/ weekend*	*Evening*	*Night/ weekend*	*Evening*	*Night/ weekend*
0–10	0.45	0.32	0.38	0.26	0.31	0.21
11–16	0.57	0.40	0.50	0.33	0.45	0.31
17–22	0.68	0.48	0.60	0.40	0.52	0.35
23–30	0.87	0.61	0.70	0.47	0.66	0.45
31–40	0.94	0.66	0.77	0.52	0.74	0.50
41–55	0.98	0.70	0.84	0.56	0.80	0.55
56–70	1.06	0.75	0.89	0.61	0.92	0.63
71–124	1.09	0.77	0.93	0.64	1.01	0.69
Number of states	8	8	26	26	5	5

Source: NARUC; *State Telephone Regulation Report*.

a. No data reported for nine states.

Table 6-8. Canadian Interprovincial Daytime Message Telephone Rates, 1980 and 1994

U.S.$ per five-minute call[a]

Mileage band	*1980 Average*	*1994 Average*	*Percent change*	*1994 U.S. rates (AT&T)*
0–8	0.51	0.51	0.0	1.15
9–20	0.72	0.73	+1.4	1.15
21–36	0.97	0.95	−2.1	1.19
37–56	1.26	1.17	−7.1	1.20
57–80	1.55	1.35	−12.9	1.25
81–110	1.80	1.43	−20.6	1.25
111–114	2.05	1.46	−28.8	1.25
145–180	2.30	1.50	−34.8	1.25
181–228	2.55	1.54	−39.6	1.25
229–290	2.79	1.57	−43.7	1.25
291–400	3.04	1.61	−47.0	1.30
401–540	3.26	1.65	−49.4	1.34
541–680	3.47	1.65	−52.4	1.35
681–920	3.69	1.68	−54.5	1.35
921–1200	3.86	1.68	−56.5	1.35

Sources: FCC, *Monitoring Report*, May 1994, p. 382; Stentor tariff, June 1994.

a. All rates are in nominal dollars, converted at the 1994 U.S./Canadian exchange rate. Canadian rates for 1994 are for calls between a rate center in Bell Canada, Island Tel, MT&T, and Newfoundland Tel and another part of Canada, excluding the Yukon Territory, Northwest Territories, and northern British Columbia.

An Empirical Analysis of Local and Intrastate (Intraprovincial) Rates

The tabular results shown above strongly suggest that the recent changes in regulatory policy in the various U.S. states have had rather limited effects on the telephone rate structure and that cross subsidies loom even larger in Canada than in the United States. We now turn to a more comprehensive empirical analysis of telephone rates that tests for the effects of differences in costs, political structure, and regulatory policy on local, long-distance, and carrier-access rates in the United States and Canada.

In our discussion of regulatory objectives in chapter 4, we derived the determinants of the equilibrium rate structure under the polar assumptions that regulators are either economic welfare maximizers or practitioners of residual pricing. Under the former assumption, rates for local connections should be related to non-traffic-sensitive costs per line in inverse relation to the absolute value of price elasticities of demand. This would suggest that rates should be higher in rural areas than in urban areas, given the greater loop lengths in rural areas and the competition from cel-

Table 6-9. Canadian Intraprovincial Daytime Message Telephone Rates, 1994

U.S.$ per five-minute call[a]

Mileage band	*Alberta*	*British Columbia*	*Manitoba*	*New Brunswick*	*New-foundland*	*Nova Scotia*	*Ontario-Quebec*	*Prince Edward Island*	*Saskatchewan*	*U.S. average intra-LATA, 12/1993 (47 states)*	*U.S. average inter-LATA, 1993 (30 states)*
0–10	0.55	0.73	0.51	0.51	0.70	0.44	0.48	0.51	0.77	0.57	0.74
11–16	0.62	0.82	0.56	0.61	0.70	0.44	0.71	0.66	0.77	0.74	0.84
17–22	0.66	0.90	0.64	0.75	0.88	0.80	0.81	0.83	0.82	0.88	0.91
23–30	0.73	0.98	0.77	0.92	0.97	0.95	0.95	1.02	0.92	1.05	1.04
31–40	0.81	1.06	0.89	1.01	1.10	1.13	1.13	1.27	1.06	1.15	1.08
41–55	0.91	1.07	0.99	1.16	1.21	1.39	1.25	1.35	1.18	1.26	1.12
56–70	0.99	1.10	1.10	1.31	1.21	1.76	1.31	1.35	1.24	1.36	1.23
71–124	1.17	1.16	1.21	1.43	1.36	1.97	1.34	n.a.	1.28	1.43	1.27
125–196	1.21	1.23	1.29	1.51	1.52	2.04	1.39	n.a.	1.33	1.56	1.35
197–292	1.24	1.28	1.38	1.57	1.67	2.16	1.41	n.a.	1.39	1.62	1.37

Sources: NBTel, General Tariff, August 1993, approved in CRTC Telecom Order 93-681; Maritime Tel and Tel Company, Limited, General Tariff, April 1992, approved in CRTC Telecom Order 92-157; Island Telephone Company, General Tariff, February 1992, Telecom Order 92-102; 1994 Saskatchewan Telephone Directory, rates effective March 1993.

a. All rates are converted at the 1994 average exchange rate of $1.3659 Canadian = U.S. $1.

n.a. Not applicable.

lular, metropolitan fiber networks and other emerging technologies in large cities. On the other hand, rates for use—toll and carrier access—should be unrelated to non-traffic-sensitive costs per line, properly measured.

Residual pricing requires that rates for all nonresidual services be set at monopoly levels. Thus for these services the only differences between residual pricing and pricing to maximize economic welfare lie in the Ramsey numbers (k). Under residual pricing, k is equal to unity; under welfare maximization k is greater than zero but less than one. However, under residual pricing, there is no simple relationship between the residual service categories—presumably, residential local service—and non-traffic-sensitive costs.

We may therefore undertake to test these two theories by positing an empirical relationship between rates and costs, demographic forces, and political variables of the following form:

$$\text{(6-1)} \qquad P_{ijt} = f(C_{jt}, DEM_{jt}, POL_{jt}, REG_{jt}, DRBOC_j, D_t),$$

in which the P_{ijt} are rates for the ith service or interexchange mileage band in the jth state (or province) in year t, C_{jt} are the cost variables in the jth state in year t, DEM_{jt} are demographic influences in the jth state in year t that affect political outcomes, POL_{jt} are structural political forces in the jth state in year t, REG_{jt} are dummy variables reflecting the regulatory frameworks adopted by telephone regulators in each jurisdiction in year t, $DRBOC_j$ is a dummy variable for the RBOC that operates in the jth state, and D_t are a set of dummy variables for each year. Note that we do not attempt to model the effects of competition in the Cournot framework of chapter 5 because we do not have a variable that provides for exogenous shifts in a major cost element to include in the analysis. Because we are now examining the rates of the LECs, access charges are output prices, not exogenous costs.

We estimate (equation 6-1) with state and provincial data, using a panel of U.S. data collected from the forty-eight mainland states and provincial data from ten Canadian jurisdictions. The estimation includes ordinary least squares for each rate category alone as well as seemingly unrelated regression techniques that allow for the interrelationships among rates resulting from regulatory tradeoffs to satisfy the carriers' revenue requirements. Thus we estimate the determinants of local business rates, local residential rates, and toll rates together.

The data are drawn from all or part of the 1987–93 period. Regulatory reform measures did not begin to spread widely among the states in the United States until 1987. We estimate equation 6-1 for annual cross sections and for pooled cross-sectional, time-series data for the entire period. Since we only have 1990 data for Canada, we only estimate the equation for a single 1990 cross section of U.S.-Canada data.

Residential and business rates are once again measured for the largest and smallest exchanges in each state. Intrastate (or intraprovincial) rates are five-minute daytime rates for three discrete mileage bands—11–16 miles, 31–40 miles, and 71–124 miles.[20] The intra-LATA rates are reported by Bell operating companies to the National Association of Regulatory Utility Commissioners (NARUC). The Canadian data are obtained from the individual companies' tariffs for 1990.

Cost conditions include the non-traffic-sensitive costs—or "revenue requirement"—per line (NTS/L) and the number of long-distance minutes per line, MIN/L.[21] NTS/L is the accounting measure that most closely captures the fixed or non-traffic-sensitive costs, while minutes per line provide a rough approximation of the effect of use on network costs. The theory of optimal or residual pricing relates rates to *incremental* NTS cost, not average NTS levels, but we have no data on incremental costs by state. It is possible that large values of NTS/L reflect the high cost of an additional line in rural areas but not in large cities.[22]

20. The choice of mileage bands does not affect the results obtained below. These three bands are chosen for economy and because the shorter mileage bands account for the bulk of intra-LATA revenues.

21. The data on long-distance minutes are far from satisfactory because of changes in the manner in which calls are originated and terminated. Some customers use dedicated circuits to deliver their calls to a carrier, and these minutes are not recorded. Moreover, the data on minutes of use that are reported by carriers are of decreasing importance in regulatory decisionmaking and may therefore be less and less accurate over time. For intrastate calling, we use intrastate dial-equipment minutes (DEMs), as reported by the National Exchange Carriers Association (NECA) to the FCC. For interstate access rates, we use switched interstate access minutes—also reported by NECA to the FCC.

22. The data on NTS costs and subscriber lines are taken from the compendium

The demographic variables include income per capita (PI/CAP), the ratio of business lines to total access lines (BUS), and the proportion of the population living in rural areas (RUR). Each of these variables may affect the ability of regulators to practice residual pricing but should have little effect on welfare-maximizing rates unless they influence the relative price elasticities of demand.

The political variables are the political composition of the legislature—specifically, for the United States, the Democrats' share of the upper chamber (DEM).[23] For Canada, we use the share of NDP and Liberals in the provincial legislature as a proxy for DEM. None of the Canadian provinces elects its regulatory commission.

Finally, the regulatory variables are dummy variables that reflect the major changes that have been occurring in the United States. For local rates, there are variables that indicate if a state has imposed some variant of price caps (RATECAP), a freeze on local rates (FREEZE), some degree of revenue sharing (REVSHR), or some form of tiered deregulation of services deemed to be competitive (TIERED). All variables assume the value of unity if the regulatory regime in question is in place and zero if it is not.[24]

For intra-LATA markets, there are a host of possibilities. Like Mathios and Rogers, we use two variables to capture the regulatory decision to block resale of bulk intra-LATA service (BLOCK) and the decision to allow facilities-based competition (COMP). In addition, we include a dummy variable for rate flexibility for the local-exchange carrier (DREG).

Equation 6-1 is estimated by ordinary least squares in linear form.[25] The choice of OLSQ to estimate equation 6-1 is somewhat problematical because it is possible that the level of rates has some effect on the choice of regulatory framework. The states that have liberalized their regulation of intrastate long-distance services may

published by the FCC in its *Statistics of Communications Common Carriers*. For Canada, NTS/L is estimated as the sum of maintenance and depreciation costs per line. This is admittedly a crude approximation, but it represents a share of total accounting costs that is virtually identical to the share of NTS in total costs for U.S. companies.

23. A dummy variable reflecting the election of the state's regulatory commission (SCOM) was also used, but its coefficients were universally insignificant.

24. No Canadian province has adopted any of these reforms.

25. Mathios and Rogers used a double log specification, but the choice between a linear and double log specification has little qualitative effect on the results.

have had lower rates before divestiture than those states that have clung to traditional regulation and limitations on intra-LATA competition. Thus one might at least explore a simultaneous-equation model of rates and regulatory regimes. As an alternative, following Mathios and Rogers, in some versions we simply include the predivestiture long-distance rate, P83, as an independent variable in the long-distance regressions. In the local-rate regressions, we use the 1980 local rates, P80, for the same purpose.

Local Rates

We begin with the U.S. rates only because Canada did not alter its regulatory framework for local-exchange carriers in the 1987–93 period.[26] The pooled cross-sectional, time-series results for each of the four local rates are shown in table 6-10.[27] Only residential rates are systematically related to non-traffic-sensitive costs per line (NTS/L), the principal regulatory measure of costs.[28] Business rates are not related to NTS/L. A large share of business lines in a state is generally conducive to lower flat rates, and higher per-capita incomes are only weakly associated with lower rates. Interestingly, a large share of rural residents leads only to higher urban business rates. And a Democratic legislature appears to translate into substantially higher flat rates.

The effects of the various regulatory reform dummies are rather surprising. Rate caps are consistently associated with much lower rates, reflecting reductions of about $1.40 per month (or about 10 to 12 percent) for residential rates and between $4.34 and $7.00 per month (or about 20 percent) for business rates. Revenue sharing and rate freezes, on the other hand, are associated with higher rates, although this effect is often not statistically significant. Tiered deregulation and full deregulation (Nebraska) are clearly associ-

26. Major regulatory changes for local-exchange and long-distance carriers did not occur until the CRTC's Decision #94-19 of September 1994. Hence, there is little to be learned about regulatory reform from including Canadian rates over the 1987–93 period.

27. The coefficients of the time dummies and the dummy variables for each RBOC are not reported since they provide no information on the effects of regulation.

28. The elasticity at the point of means is only 0.2. Thus a state with NTS costs that are 10 percent above average would generally be predicted to have residential rates that are only 2 percent or about 20 cents higher than average, ceteris paribus.

Table 6-10. Estimates of the Determinants of U.S. BOC Local Flat Rates from Pooled Cross-Section Time-Series Data, 1987–93[a]

t Statistics in parentheses

Dependent variable: real monthly flat rate (1982–84 U.S.$)

Variable	*Urban residential*		*Rural residential*		*Urban business*		*Rural business*	
NTS/L	14.81	12.98	13.83	12.70	16.10	9.31	17.91	17.16
	(2.61)	(2.53)	(2.69)	(2.54)	(1.08)	(0.67)	(1.43)	(1.37)
PI/CAP	−0.25	−0.21	−0.22	−0.19	−0.01	−0.07	−1.26	−1.26
	(−1.90)	(−1.63)	(−1.88)	(−1.72)	(−0.01)	(−0.15)	(−4.17)	(−4.29)
BUS	−11.85	−7.32	−18.57	−8.86	−57.60	−17.07	−50.72	−27.13
	(−1.86)	(−1.27)	(−3.30)	(−1.63)	(−3.29)	(−1.20)	(−3.55)	(−1.74)
RUR	0.58	1.07	−0.79	−0.09	5.09	10.62	−2.58	−0.50
	(0.68)	(1.27)	(−1.09)	(−0.12)	(2.55)	(4.82)	(−1.34)	(−0.24)
DEM	5.57	5.44	6.20	4.82	14.98	12.19	12.94	8.62
	(5.83)	(5.77)	(6.32)	(5.14)	(5.84)	(4.77)	(5.83)	(3.91)
RATECAP	−1.40	−1.37	−1.38	−1.41	−7.00	−5.35	−4.45	−4.34
	(−3.89)	(−3.74)	(−6.32)	(−3.67)	(−5.00)	(−4.54)	(−4.18)	(−4.25)
REVSHR	0.46	0.33	0.12	0.01	2.42	1.40	0.94	0.78
	(1.69)	(1.19)	(0.50)	(0.03)	(2.61)	(1.86)	(1.54)	(1.33)
TIERED	0.57	0.54	0.71	0.88	0.13	−0.18	1.34	1.41
	(2.23)	(2.17)	(2.85)	(3.49)	(0.15)	(−0.23)	(2.20)	(2.37)
FREEZE	0.93	1.22	0.89	0.79	1.80	1.62	0.95	0.70
	(2.41)	(2.93)	(2.28)	(2.10)	(1.45)	(1.30)	(1.27)	(0.98)
DEREG	2.39	2.37	3.43	3.77	2.43	2.81	9.45	9.81
	(6.09)	(6.32)	(9.25)	(9.96)	(2.38)	(2.92)	(9.60)	(10.26)
P80	—	0.08	—	0.63	—	0.73	—	0.53
		(2.53)		(4.36)		(5.96)		(3.90)
R^2	0.49	0.50	0.46	0.48	0.51	0.59	0.56	0.58

a. Coefficients of time dummies and dummy variables for each RBOC are not reported.

Table 6-11. Ordinary Least Squares Estimates of the Determinants of Local Flat Rates in the United States and Canada, 1990

t Statistics in parentheses

Dependent variable: real monthly flat rate (1982–84 $ per month)

	Residential		*Business*	
Variable	*Urban*	*Rural*	*Urban*	*Rural*
NTS/L	0.015	−0.0063	0.030	0.019
	(1.82)	(−0.87)	(1.09)	(0.00)
PI/CAP	0.19	−0.17	0.69	−0.26
	(1.16)	(−1.12)	(1.20)	(−0.48)
BUS	−0.29	−0.15	−0.76	−0.73
	(−2.21)	(−1.33)	(−1.65)	(−1.70)
RUR	0.029	0.026	0.069	0.045
	(1.29)	(1.29)	(0.92)	(0.62)
DEM	6.48	2.83	23.02	13.20
	(3.29)	(1.60)	(3.31)	(2.00)
CANADA	−7.48	−7.34	−12.63	−18.81
	(−6.31)	(−6.93)	(−3.09)	(−4.78)
R^2	0.631	0.666	0.391	0.532

ated with higher rates. The Nebraska experience in deregulation is particularly instructive. Urban business rates were increased the least (about 8 percent) while rural rates are much higher, ceteris paribus, than rates in other states—as much as 28 percent higher.

Thus full deregulation and rate caps work decidedly in opposite directions. Rate caps are associated with larger reductions in business rates than in residential rates; Nebraska's full deregulation—protected by entry restrictions—has led to larger increases in rural rates than in urban rates. Nebraska's experience is that rural business rates have risen much more than rural residential rates, a result that may or may not be consistent with Ramsey pricing. Otherwise, regulatory reforms, such as revenue sharing, have had remarkably small effects on individual rates or the structure of rates.

To estimate the effects of the demographic, regulatory, and political variables for the larger sample of U.S. and Canadian rates, we add a dummy variable for Canada (CANADA) and estimate the equation for 1990 alone (table 6-11). As before, Democratic or left-liberal legislatures are associated with higher local rates. The coefficients of BUS are negative, but statistically significant only for urban residential rates. Of perhaps greater interest is the fact that Canadian flat rates were significantly lower—from $7.34 to $18.81 per month—than their U.S. equivalents in 1990, after allowing for

Table 6-12. Estimates of the Determinants of U.S. BOC Intra-LATA MTS Rates and Intrastate Access Rates from Pooled Cross-Section Time-Series Data, 1987–93[a]

t Statistics in parentheses

Dependent variable: real rate for a five-minute call

Variable	*11–16 miles*		*23–30 miles*		*41–55 miles*		*Access rate*
NTS/L	0.86	0.67	2.06	1.24	1.62	1.43	0.006
	(4.22)	(3.82)	(6.22)	(4.54)	(4.27)	(4.43)	(0.24)
PI/CAP	−0.01	−0.01	−0.01	−0.01	−0.02	−0.01	−0.004
	(−1.83)	(−1.84)	(−0.92)	(−0.83)	(−1.49)	(−1.23)	(−0.45)
BUS	−2.45	−0.47	−3.19	−0.78	−1.01	−1.07	−0.007
	(−7.97)	(−2.25)	(−8.74)	(−2.92)	(−2.06)	(−2.47)	(−0.20)
RUR	−0.08	−0.02	−0.10	−0.01	0.12	−0.02	0.004
	(−1.75)	(−0.68)	(−1.58)	(−0.14)	(1.60)	(−0.27)	(0.60)
DEM	0.11	0.10	0.22	0.21	0.45	0.47	−0.01
	(2.33)	(3.23)	(3.38)	(4.53)	(5.31)	(6.26)	(−1.66)
MIN/L	−0.04	0.01	−0.04	0.01	−0.02	−0.01	−0.002
	(−4.69)	(2.00)	(−3.79)	(1.45)	(−1.37)	(−0.04)	(−2.85)
COMP	−0.02	−0.02	0.02	−0.01	−0.01	−0.04	−0.003
	(−1.12)	(−1.48)	(1.14)	(−0.39)	(−0.27)	(−1.70)	(2.12)
BLOCK	0.01	0.02	0.05	0.04	0.06	0.10	−0.007
	(0.34)	(1.23)	(1.50)	(1.92)	(1.74)	(3.39)	(−2.50)
DREG	0.09	0.08	0.19	0.13	0.16	0.14	0.005
	(3.82)	(7.04)	(6.98)	(7.29)	(4.65)	(4.73)	(3.11)
P83	...	0.77	...	0.72	...	0.47	...
		(18.45)		(15.03)		(9.01)	
R^2	0.58	0.81	0.65	0.80	0.60	0.70	0.66

a. Coefficients of time dummies and RBOC dummy variables not reported.

differences in cost per line and demographics.[29] Otherwise, few insights are to be gleaned from combining U.S. and Canadian data because Canada has not yet implemented any of the regulatory-reform measures.

Long-Distance Rates

The results for the pooled time-series cross-sectional estimates of the determinants of U.S. intra-LATA long-distance rates are shown in table 6-12.[30] These results provide a number of insights into both the politics and economics of postdivestiture ratemaking.

First, non-traffic-sensitive costs have a surprisingly significant association with rates in all three mileage bands shown, suggesting that regulators in states with high local-access costs use long-distance rates to cover these costs. The elasticity of intra-LATA rates with respect to NTS/L is between 0.25 and 0.47, about double the elasticity for residential flat rates.

Second, the number of minutes per line is only weakly associated with intra-LATA rates. The lack of significance of use in this equation may simply reflect an identification problem, confounding the effects of regulatory design and market demand for intra-LATA use.

Finally, and most important, is the result that competition is associated with lower rates, but most of the coefficients of COMP are not statistically significant. However, regulatory flexibility is associated with higher rates, and the result is statistically significant—a result consistent with our earlier tabular representations of the data. The pooled regressions also suggest that blocking of resale is generally associated weakly with higher rates, a result that is irrelevant since no state continues to block resale. Thus competition appears to have little association with lower rates, but rate flexibility is associated with rates that are 20 to 30 percent higher.

29. Equations with SCOM and the regulatory flexibility variables are not reported because they add nothing to the explanatory power of the equation.

30. The three bands chosen are the second, fourth, and sixth of the ten bands used by Mathios and Rogers. The results for the other bands are qualitatively similar and are therefore not reported; moreover, there is a substantial loss of degrees of freedom in the longer mileage bands because of the number of small states in the eastern United States.

Table 6-13. Ordinary Least Squares Estimates of the Determinants of U.S. and Canadian Intra-LATA (Intraprovincial) Daytime Toll Rates, 1990
t Statistics in parentheses
Dependent variable: price per call (1982–84 U.S.$ per five minutes)

Variable	*11–16 miles*	*23–30 miles*	*41–55 miles*
P80	0.52	0.36	0.38
	(3.75)	(2.63)	(2.82)
NTS/L	0.00058	0.0013	0.00082
	(0.99)	(2.03)	(1.08)
PI/CAP	0.015	0.0053	0.0010
	(1.22)	(0.39)	(0.06)
PBUS	−0.025	−0.17	−0.0077
	(2.65)	(−1.60)	(−0.63)
PRUR	0.00050	0.0013	0.0039
	(0.30)	(0.69)	(1.81)
DEM	0.13	0.43	0.72
	(0.86)	(2.51)	(3.52)
CANADA	−0.03	−0.18	−0.26
	(−0.28)	(−1.56)	(−1.96)
BLOCK	0.19	0.24	0.34
	(2.31)	(2.57)	(3.13)
COMP	0.08	0.11	0.18
	(1.34)	(1.57)	(2.20)
DREG	0.12	0.32	0.20
	(1.46)	(3.37)	(1.81)
R^2	0.410	0.541	0.530

Every local-exchange company charges long-distance carriers an access rate for originating or terminating their intrastate calls. In 1993 these state access charges accounted for nearly 8 percent of U.S. local carriers' revenues.[31] State access charges vary inversely with minutes of use, but the effect of non-traffic-sensitive costs (NTS/L) is not statistically significant. Once again, full regulatory flexibility in intra-LATA rates leads to higher carrier access rates. In addition, intra-LATA competition appears to have only a mild inverse association with access rates with an elasticity of only −0.08.

The addition of the eight Canadian observations to the intra-LATA regressions for 1990 (table 6-13) does not yield any new insights. The dummy variable for Canada (CANADA) has negative

31. The access rates we use in this chapter have been described in chapter 5.

coefficients that increase in absolute value with mileage, suggesting that Canadian intraprovincial rates are lower than U.S. intra-LATA rates in the longer mileage bands. Otherwise, the results are little changed by the addition of Canada to the sample.[32] Blocking resale raises rates; rate flexibility raises the rates. Democratic (and liberal-left) legislatures are associated with higher rates, but rates decline with an increase in the share of business lines in total access lines in the shorter mileage bands.

These results cast doubt on the notion that regulatory reform in the states that simply takes the form of allowing carriers greater rate flexibility will by itself lead to lower long-distance rates and more efficiently structured local rates. As long as regulatory commissions regulate access charges for competitive carriers, they can achieve any rate structure they desire regardless of the degree of competition that they allow. All else equal, competition leads to lower rates, but this effect of competition is quite muted. In the shortest intra-LATA mileage bands, the cost of local access and switching surely composes most of the cost of the call. As long as the regulated local-exchange carriers control this access, it would appear that their costs and the regulation of access—not intra-LATA competition—will dominate the structure of intra-LATA rates.

Seemingly Unrelated Regressions

The errors across the four flat-rate equations (urban and rural; business and residential), the three intra-LATA equations, and the access-rate equation are likely to be correlated because a regulator's (or regulated firm's) choice to recover a disproportionately large share of costs from one service should lead to a correspondingly lower charge for some other service, given NTS costs. Therefore, we estimate all of the rate equations together, using seemingly unrelated regression (SUR) techniques.

We do not report all of the estimated coefficients from this exercise because the results largely reinforce the results reported in tables 6-10 and 6-12. Instead, we show the effects of various

32. A precise statistical test is not available given the number of independent variables and only eight Canadian observations.

Table 6-14. Estimates of the Effect of Various Reform Measures on Local Rates from Seemingly Unrelated Regressions, 1987–93
$ per month, flat rate

		Residential		*Business*	
Reform measure	*Year*	*Urban*	*Rural*	*Urban*	*Rural*
Rate caps	1987	n.a.	n.a.	n.a.	n.a.
	1988	n.a.	n.a.	n.a.	n.a.
	1989	−3.38[a]	−3.64[a]	−9.30[a]	−7.44[a]
	1990	−2.60[a]	−2.65[a]	−5.77[a]	−5.44[a]
	1991	−1.31[a]	−2.09[a]	−1.58	−6.11[a]
	1992	−2.09[a]	−2.83[a]	−6.56[a]	−8.52[a]
	1993	−1.89[a]	−2.72[a]	−6.44[a]	−8.01[a]
Profit sharing	1987	+3.40[a]	+3.16[a]	+12.93[a]	+8.42[a]
	1988	+1.27	+1.13	+5.37	+2.81
	1989	+0.30	−0.17	+0.27	−1.68
	1990	+0.49	+0.62	+0.10	−1.20
	1991	−0.87	−1.06[a]	−1.37	−4.12[a]
	1992	−0.34	−0.93[a]	+0.59	−2.71[a]
	1993	+1.13[a]	+0.51	+3.68[a]	+0.46
Tiered deregulation	1987	−1.94[a]	−2.30[a]	−8.86[a]	−8.60[a]
	1988	+1.61[a]	+0.48	−0.02	−0.75
	1989	+1.43[a]	+0.71	+3.38[a]	+1.11
	1990	+1.12[a]	+1.11[a]	+4.14[a]	+4.88[a]
	1991	+0.49	+0.30	+0.84	+0.18
	1992	+0.35	−0.44	+0.24	−2.66[a]
	1993	+1.07[a]	+0.51	−1.65	−1.31
Rate freeze	1987	n.a.	n.a.	n.a.	n.a.
	1988	−1.03	−1.95	−12.08[a]	−6.01
	1989	−0.06	−1.43	−7.15[a]	−2.16
	1990	+3.42[a]	+3.04[a]	+1.03	+0.16
	1991	+2.62[a]	+0.96	−2.70	+0.48
	1992	+0.01	+1.04	−5.61[a]	+0.28
	1993	+0.09	+0.99	−6.75[a]	−1.20

a. Statistically siginificant at the 5 percent confidence level.
n.a. Not applicable.

changes in regulatory regime over the 1987–92 period in tables 6-14 and 6-15.[33]

The effects of regulatory reform on local rates (table 6-14) provide a few additional insights. First, rate caps are uniformly associated with lower rates and with a reduction in the ratio of business to residential rates. Second, profit (or revenue) sharing was asso-

33. We do not report the coefficients of BLOCK because the blocking of intra-LATA resale has disappeared in the United States.

Table 6-15. Estimates of the Effect of Competition and Rate Flexibility on Intra-LATA Rates from Seemingly Unrelated Regressions, 1987–93
$ per five-minute daytime call

Reform measure	*Year*	*11–16 miles*	*23–30 miles*	*41–55 miles*
Competition	1987	−0.05	+0.07	+0.04
	1988	+0.18[a]	+0.31[a]	+0.27[a]
	1989	+0.05	+0.16	+0.22[a]
	1990	+0.04	+0.10	+0.14
	1991	−0.04	−0.02	+0.10
	1992	−0.18[a]	−0.18[a]	−0.17[a]
	1993	−0.16[a]	−0.10[a]	−0.01
Rate flexibility	1987	+0.27[a]	+0.37[a]	+0.17[a]
	1988	−0.03	+0.24[a]	+0.15
	1989	+0.06	+0.31[a]	+0.21[a]
	1990	+0.09	+0.27[a]	+0.16
	1991	+0.08	+0.25[a]	+0.17
	1992	+0.08	+0.19[a]	+0.12
	1993	+0.10[a]	+0.16[a]	+0.22[a]

a. Statistically significant at the 5 percent confidence level.

ciated with higher rates in 1987, but the coefficients of REVSHR become negative and generally significant in 1991–92. Selective or tiered deregulation is associated with lower rates in 1987, but this effect atrophies and actually reverses in 1989–90 before receding into general insignificance. Finally, rate freezes are not generally associated with lower rates, a result that may seem counterintuitive to those unfamiliar with the generally declining costs of telephony.

The SUR regression results on long-distance rates (table 6-15) help to explain the lack of significance of the competition (COMP) variable in table 6-12. Competition does not appear to lead to lower rates until 1992. In fact, competition is associated with higher rates in 1988, perhaps because the states that admitted competition this soon were states with the highest rates. On the other hand, rate flexibility is consistently associated with higher intra-LATA rates.

Although we do not report the coefficients of NTS/L, we should note that they are generally statistically insignificant, particularly after 1988. Rates appear to move without regard to the underlying revenue requirement, driven by political considerations, demographics, and the inexorable pressure of liberalization in densely populated areas.

We conclude from this analysis that rate caps have been part of a general policy to reduce local rates and to rebalance them marginally toward relative costs. Competition is only recently having a

Table 6-16. A Comparison of Ordinary Daytime MTS Rates across Regulatory Jurisdictions, 1993–94
$ per five-minute call

Mileage band	*Intra-LATA RBOCs (December 1993)*	*Intrastate inter-LATA AT&T (mid-1993)*	*Interstate AT&T (April 1993)*
0–10	0.57	0.74	1.05
11–16	0.74	0.84	1.10
17–22	0.88	0.91	1.10
23–30	1.05	1.04	1.10
31–40	1.15	1.08	1.10
41–55	1.26	1.12	1.10
56–70	1.36	1.23	1.15
71–124	1.43	1.27	1.15
125–196	1.56	1.35	1.15
197–292	1.62	1.37	1.15

Source: National Association of Regulatory Commissioners, AT&T, FCC, *Monitoring Report*.

major effect on intra-LATA rates, but rate flexibility for local-exchange carriers that are not required to provide 1+ intra-LATA access has consistently perverse effects—namely, it raises intra-LATA rates. Thus, while intra-LATA rates are falling relative to local access rates, this decline cannot be attributed to the liberalization efforts of individual state regulatory commissions. Nor may we conclude that regulatory reform in the states has moved very far down the path of rational rate setting.

We previously concluded that inter-LATA (and interstate) competition has been responsible for a certain amount of rate reduction and narrowing of the variance of rates across states. To wrap up our assessment of the effectiveness of changes in intrastate regulation, including the modest liberalization that has occurred, we provide a comparison of ordinary MTS daytime rates in the three U.S. regulatory jurisdictions—intra-LATA, intrastate inter-LATA, and interstate—in 1993, the most recent year for which we have complete intra-LATA data (table 6-16). These data show that very short intrastate calls are still priced below the interstate rates for similar calls. Indeed, the sharp taper in rates from the shortest to the longest in our intra-LATA tabulation is undoubtedly not explained by cost differences. In the longer mileage bands, the intra-LATA rates are substantially above the inter-LATA and interstate rates, suggesting less competition in the intra-LATA jurisdiction. Moreover, the intrastate inter-LATA rates are also substantially above the interstate rates in the longer mileage bands, thus suggesting that there is room for further competition in the intrastate inter-LATA market.

Finally, it should be noted that the rates in table 6-16 reflect the fact of regulation. In every case, a carrier must maintain uniform rates for calls of similar distance despite very large differences in volume. In some states, longer calls may be on only low-density routes, while AT&T's interstate rates obviously reflect a very large number of diverse routes for each mileage band. In short, even if rates are reasonably competitive by some measure under today's regulatory regimes, they would undoubtedly change, perhaps dramatically, if deregulated competition were to be substituted for the current regulatory regimes.

Rate Rebalancing and Competition in the Local Loop

Our results cast substantial doubt on the proposition that states are moving rapidly toward liberalization of intrastate telephone services. Local access rates in the United States are still woefully distorted; rural rates are still lower than urban rates despite the much higher cost of incremental service in less populous areas. Business (non-Centrex) rates are still more than double residential rates despite relatively similar costs. Until 1995 state regulators have steadfastly resisted requiring dialing parity for competitors in intra-LATA toll markets so as to generate subsidies for the underpriced local-access services.

Twelve years after the breakup of AT&T, the divested RBOCs and the long-distance companies continue to debate whether actual competition in the local loop must exist before any easing of the inter-LATA line-of-business restriction. Unfortunately, little attention is given to the fact that competition will immediately undercut the distorted rate structure erected by state regulatory officials. The RBOCs are in the difficult position of being forced to open up their local networks to competition without being allowed to reprice services immediately. If competition develops under these conditions, the RBOCs (and other local-exchange companies) will suffer an immediate loss of their high-margin business while being saddled with the nonremunerative services. As this book is being written, the last vestiges of protection in the intra-LATA mar-

Table 6-17. Growth in U.S. Local Carriers' Real Local Monthly Rates and Local Revenues, 1987–93
1987 $ per month

Rate/revenues	*1987*	*1993*[a]	*Annual rate of growth*
Urban residential[b]	16.75	13.86	−3.2
Rural residential[b]	13.55	11.60	−2.6
Urban business[c]	43.63	33.35	−4.5
Rural business[c]	32.69	25.64	−4.0
Total local revenues per switched line	27.58	22.83	−3.1

Sources: NARUC; FCC, *Statistics of Communications Common Carriers.*
a. The data for 1993 are deflated by the CPI-U = 1.00 in 1987.
b. Forty-two states.
c. Thirty-six states.

ket are falling—in other words, states are finally moving to implement 1+ intra-LATA access. If the states do not act to rebalance rates (or impose large intra-LATA access charges) as they liberalize intra-LATA services, they will severely impair the local-exchange companies.

The paltry effects of regulatory reform on local-access carriers may be seen in table 6-17. Using our data on local rates for those jurisdictions for which we have complete data in 1987 and 1993, we find that inflation-adjusted residential rates have fallen at a rate of between 2.6 percent and 3.2 percent per year since 1987. Flat business rates have fallen at a rate of between 4.0 and 4.5 percent per year. These declines probably bracket the rate at which productivity has increased in the industry. In other words, real rates have fallen at a rate that reflects technical progress. Similarly, real local revenues per switched line have fallen at an annual rate of 3.1 percent. By contrast, real intra-LATA rates have fallen at a 4.5 percent rate over the past decade (table 5-2). Since efficient pricing requires that local rates increase substantially relative to long-distance rates, this period of experimentation with regulatory reform and rate flexibility cannot be said to have rid the industry of its burden of regulatory rate distortions.

Ironically, opponents of RBOC entry into inter-LATA markets can use state regulatory commission intransigence in rebalancing rates as their rationale for continuing the quarantine. As long as the regulated subscriber rates in less-dense areas, particularly for residences, are below the prospective entrants' long-run incremental cost of serving them, entry will be limited to dense areas with a heavy concentration on business subscribers. Competition will not be ubiquitous unless there is a technological revolution that bankrupts the wire-line carriers; therefore, there will always be a case for maintaining the quarantine because of state regulation.

Incentive Regulation and the Diffusion of New Technology

One of the motivating factors for switching from traditional to incentive regulation is to encourage more rapid introduction of cost-saving technology or new services. Because carriers are able to retain more of the gains from such innovations under incentive regulation, we should be able to observe higher rates of diffusion of important new technologies in those local-exchange companies regulated by more incentive-compatible regimes than in those under traditional cost-based (ROR) regulation. Indeed, Taylor and associates obtain precisely this result when they examine modernization data for LECs from the 1980–94 period, using a fixed effects model in which the only structural variables are regulatory-reform variables: banded ROR, ROR with sharing, flexible pricing ("tiered deregulation" in our parlance), price caps, and social contracts.[34] Their dependent variable is the rate of diffusion of new technologies—digital switching, fiber-optic transmission, ISDN, and SS7 signaling. The 1980–89 data are those reported by the carriers to the FCC; the 1990–94 data are forecasts by the firms in 1990. We cannot know whether these latter estimates were actually realized, nor whether they drive the results.

Taylor and his colleagues find that incentive regulation accelerates digital switching and fiber deployment by about eleven months; ISDN and SS7 by five months. Tiered deregulation and banded ROR appear to have the most systematic positive effects on diffusion. More recently, Greenstein, McMaster, and Spiller have examined the effect of incentive regulation on the deployment of these technologies in the 1987–91 period.[35] They find that rate caps

34. William E. Taylor, Charles J. Zarkadas, and J. Douglas Zona, "Incentive Regulation and the Diffusion of New Technology in Telecommunications," International Telecommunications Society, Ninth Annual Conference, June 1992.

35. Shane Greenstein, Susan McMaster, and Pablo T. Spiller, "The Effect of Incentive Regulation on Local Exchange Companies' Deployment of Digital Infrastructure," paper presented at AEI Telecommunications Summit, July 7, 1994.

increase the deployment of all four technologies by as much as 127 percent, but that a combination of rate caps and earnings sharing or simply earnings sharing alone has smaller, but nonetheless significantly positive, effects on BOC investments.[36] We have attempted a more modest confirmation of their findings by examining the amount of fiber optics per line and the number of ISDN circuits per line in the lower forty-eight U.S. states in 1991. Consistent data on these or other forms of new technology are simply not available on a state basis for sufficient years to estimate a fixed-effects model. We find that ISDN diffusion is positively associated with tiered regulation and (Nebraska) deregulation, but not with rate caps and revenue sharing, a result similar to that obtained by Taylor and his colleagues but different from the Greenstein and associates study. For fiber penetration, however, the results are similar in sign but not statistically significant.[37]

Conclusion

While the FCC has, under its regulatory authority, reduced the distortions in pricing access and interstate toll, state authorities have only recently begun to move. Intrastate toll rates have not fallen at the same pace as interstate rates. Surprisingly, there is evidence that intra-LATA toll rates are generally not lower in those states that have liberalized intra-LATA service than in those that have not, and they are also generally lower in states that continue to regulate rates than in those that allow full rate flexibility. Our more detailed empirical results generally confirm the perverse effects of rate flexibility, but our results provide only mild support for the notion that intra-LATA competition, without equal access 1+ dialing, reduces rates.

Our results strongly confirm that Canadian local rates are lower than their U.S. counterparts. Surprisingly, even Canadian intra-

36. The effects for GTE companies in their sample are far less favorable, a result that they ascribe to the fact that GTE owns a manufacturing subsidiary.

37. The estimates are OLSQ, using dummy variables for RBOCs and demographic variables, such as the share of business lines and the share of population in rural areas. Surprisingly, rural population is *directly* associated with fiber diffusion, but the share of business lines is not significant in either regression.

provincial toll rates are also lower than U.S. intra-LATA MTS rates. The Canadians offset this deficiency in rates with sharply higher interprovincial rates.

Equally disturbing is the tendency of the U.S. state regulators and Canadian authorities to cling tenaciously to their policies of charging urban consumers more than rural consumers for telephone service and businesses far more than residences for similar flat-rate service. The regulatory reform measures, such as rate caps or revenue sharing, have very little effect on these local-rate distortions. These distortions will prove important in our further discussion of regulatory reforms, for they provide the incentive for organized interest groups to resist changes in the current regulatory structure.

Chapter Seven

Local-Access Competition from Other Technologies

CONTINUED REGULATION of telephony is premised on the existence of a natural monopoly or bottleneck in local-access service. However, this natural monopoly is now or soon will be under attack from a number of new technologies, including cellular telephony, metropolitan fiber-optics systems (competitive-access providers, or CAPs), personal communications systems (PCS), cable television systems, and other forms of fixed wireless technologies. A few of these technologies already exist; the latter two are in development. However, no one has yet made a major frontal assault on local telephone companies. As a result, there is no market evidence of the contestability of the local loop.[1]

The local telephone market is about to change, perhaps dramatically. The protection state regulatory commissions provide to their franchised telephone carriers is weakening. Large joint ventures are being formed by cable television companies and telephone companies outside the franchise areas of the latter to build switched capacity, perhaps through PCS, into existing cable plant. The costs of wireless and fiber optic/coaxial cable technologies are falling dramatically and will likely fall even more rapidly once these systems are installed and the competitive war for local subscribers begins in earnest.

In this chapter, we examine the prospective costs of alternative local-loop technologies and the prospects for local competition in

1. A recent study by Economics and Technology, Inc., and Hatfield Associates, *The Enduring Local Bottleneck: Monopoly Power and the Local Exchange Carriers* (Boston, 1994), prepared for AT&T, MCI, and CompTel, argues that dominance and control of local access by local-exchange companies will persist for at least five to ten more years.

telephony. This is necessarily somewhat speculative but important for the discussion in chapter 8 that focuses on future regulatory strategies.

Alternative Technologies

The rapid pace of technical change in electronics and communications technologies renders any precise analysis of the costs of alternate local-loop technologies extraordinarily difficult. It is safe to say, however, that the number of potentially competitive technologies is growing rapidly and that the embedded plant of the existing companies is threatened today as never before. The most likely candidates at this juncture appear to be the following:

—Advanced cellular or PCS networks;
—Wireless fixed-access systems;
—Cable television companies (perhaps using PCS technology);
—Satellites;
—Fiber-optics networks (competitive-access providers).

There are a number of variants of several of these technologies (and even other technologies, such as a mobile system using satellites), but we will concentrate on only these five, providing cost estimates where data are available. In so doing, we rely heavily on the existing literature and confine ourselves to delivery of narrowband voice/data services.[2] The more exotic switched broadband services of the information superhighway are subject to much less certainty at this juncture.

Current Cellular Telephony

Mobile telephone service has been available since 1946. Throughout the 1950s and 1960s, the demand for mobile telephony

2. In particular, we use the studies performed by David P. Reed, former member of Office of Plans and Policy, U.S. Federal Communications Commission (now of Cable Labs, Boulder, Colo.), for several of the more promising options. One of these was "The Prospects for Competition in the Subscriber Loop: The Fiber-to-the-Neighborhood Approach," presented at the twenty-first annual Telecommunications Research Policy Conference, Baltimore, Maryland, September 1993.

increased steadily, but this growth was constrained by the lack of sufficient spectrum and the technology. Each mobile telephone had exclusive use of a frequency in the entire franchise area. In the 1960s, AT&T developed a new cellular approach to mobile telephony that allowed the same frequency to be used at lower power in a number of different locations within the same franchise area. The area would be divided into hexagonal cells that would constitute separate transmission areas. As mobile customers passed from one cell to another, they would be passed off to different fixed transmission facilities. In this way, the same frequency could be used by customers in a number of different cells in the area without causing interference with one another. As the density of use increased, the average cell size could be reduced, thereby permitting an even greater sharing of frequencies.

In 1970, the Federal Communications Commission began a proceeding to investigate the possibility of expanding the amount of U.S. spectrum dedicated to mobile services. The FCC initially identified 115 MHz of spectrum that it proposed transferring to such uses, but finding a political compromise for allocating this spectrum was difficult, even though the spectrum was largely idle at the time.[3] In 1975, the FCC finalized the transfer of this spectrum to mobile uses, but it was not until 1982 that its plan for cellular service was also finalized. Two cellular services were to be assigned 20 MHz each, one of which was to be the existing wire-line common carrier. Subsequently, each of these bands was expanded to 25 MHz. In October 1983, the first cellular system, operated by Amaurotic, began commercial operation.

Cellular service thus began as a government-licensed duopoly with licenses extending over metropolitan areas and later to rural service areas (RSAs). At first a large number of independent operators owned the non-wire-line licenses, but a series of acquisitions and mergers soon led to consolidation within the industry. By 1993, the largest ten companies—including the seven regional Bell holding companies (RHCs)—accounted for more than 80 percent of total cellular subscribers. However, because of restrictions in the

3. Most of the spectrum in question was unused UHF television broadcasting spectrum. Despite the fact that this spectrum had lain fallow for decades, broadcasters opposed transferring it to another use.

AT&T decree, the RHCs cannot operate across local access and transport areas (LATAs). These companies are confined to operating a series of local cellular systems that are interconnected through interexchange service offered by AT&T, MCI, Sprint, and the other carriers.

At first, little attention was paid to the constraints imposed by the spectrum allocated to cellular service. Analog systems were built, and analog transceivers proliferated. But as the number of subscribers increased (particularly in the largest cities), equipment manufacturers began to develop new digital technologies to allow more efficient use of the spectrum. The first of these technologies, time-division multiple access (TDMA), would approximately triple the capacity of the existing spectrum. A more recent approach, code-division multiple access (CDMA), would expand capacity of a given spectrum allocation more than tenfold. Digital cellular has been slow to develop. Perhaps this is because the capacity constraints for analog systems have not been reached in most markets, and conversion has been limited thus far to the TDMA technology.

In 1992, a new cellular company, Fleet Call (now Nextel) entered the market by purchasing specialized mobile radio (SMR) licenses and, with FCC permission, converting them to cellular use. This company and a number of others are now using smaller blocks of spectrum with TDMA technology to compete with existing cellular carriers in cities such as Los Angeles and Chicago.[4] Fleet Call's entry was significant for one important reason: it marked the first time a new competitor entered the market by offering to make better use of spectrum than what the FCC had originally envisioned. If the FCC is willing to allow competitive entry into a variety of markets through voluntary alterations of its spectrum allocation plan, various entrenched monopolies might be affected. For instance, cable television may be vulnerable to multichannel entrants in bands other than those allocated to MMDS (multichannel, multipoint distribution service, or "wireless cable," as it is popularly known).

4. As this book goes to press, Nextel is apparently having technical problems in rolling out its service. Personal communications services (discussed below) may well provide a more serious competitive threat than the SMR-based services.

Table 7-1. U.S. Cellular Market

Date (end of year)	*Subscribers (thousands)*	*Average monthly bill ($)*
1985	340	n.a.
1986	682	n.a.
1987	1,231	96.83
1988	2,069	98.02
1989	3,509	89.30
1990	5,283	80.90
1991	7,557	72.47
1992	11,033	68.68
1993	16,009	61.48
1994	19,283	56.21

Source: Cellular Telecommunications Industry Association, March 1995.
n.a. Not available.

The Economics of Cellular Service

At first, cellular service did not seem to be a serious competitor of the wire-line companies. It was in fact seen as a complementary service, funneling calls through wire-line companies to other jurisdictions. With 93 percent of U.S. households subscribing to regular wire-line service, cellular service was not purchased as a *replacement* for such service. Instead, it was marketed as a new service—either for business subscribers or for affluent households. Marketing costs were high and remain so, as cellular companies attempt to persuade consumers to purchase their service in addition to regular wire-line service.

The growth of cellular service in its first few years was so slow that many asked if it had a future as long as it was constrained by the economics of analog transmission in the existing spectrum allocations.[5] Four years after Ameritech opened the first cellular business to subscribers, only 1.2 million had signed up throughout the United States (table 7-1). At that time, the lowest cost cellular receiver was about $500. But by 1993, cellular was growing much more rapidly. Cellular receivers had dropped in price to as little as $50 for some inexpensive models and about $200 for Motorola's most popular portable set. Between December 1991 and December 1994, the number of subscribers more than tripled, increasing from 7.6 million to 24.1 million. But cellular service remains an ad-

5. George Calhoun, *Digital Cellular Radio* (Norwood, Mass.: Artech House, 1988).

junct or complement to regular wire-line service in all but a few communities.

Careful examination of the current cost structure for cellular and wire-line services suggests that their average costs may not differ greatly for their respective current average use patterns, but average minutes of use are decidedly different for the two services. Cellular is currently much more expensive to use at peak hours but is still economical for those with relatively light telephone use. This is because the fixed cost of customer connections is much higher for wire-line carriers, but peak-hour call handling capacity is more expensive for cellular carriers.

With current technology, the cost of new paired copper-wire local-exchange plant ranges from $400 to more than $2,000 for residential lines, depending on loop length.[6] Bellcore studies estimate the average cost for Bell operating companies (BOCs) to be in the $1,750–$1,800 range per copper loop,[7] although Reed estimates that these costs will fall substantially in the next few years.[8] There are no "loops" in cellular systems. Rather, these systems use cells whose number increases with traffic density. The cost of building these cells obviously depends on local site costs and traffic density. At some point, cell division becomes very expensive, and systems with analog technology must contemplate more spectrum-efficient digital technologies. However, current estimates of the embedded cost per cell is about $900 per subscriber for analog cells sufficient to accommodate 1,000 subscribers. The incremental cost of cellular plant per subscriber at today's prices may be $700 or even less,[9] except in central cities where site-acquisition costs may be high.

6. See Bridger M. Mitchell, *Incremental Costs of Telephone Access and Local Use* (RAND Corporation, July 1990), pp. 45-48.

7. George Calhoun, *Wireless Access and the Local Telephone Network* (Boston: Artech House, 1992), p. 77.

8. David P. Reed, *Residential Fiber Optic Networks: An Engineering and Economic Analysis* (Boston: Artech House, 1992), appendix B, pp. 88–91.

9. The Cellular Telecommunications Industry Association (CTIA) reports that cellular companies spent $5 billion on plant and equipment in 1994 while adding 8.1 million subscribers. Not all of this investment was on expansion, nor is it safe to assume that new plant is built precisely in anticipation of the need for accommodating new subscribers. Nevertheless, the $617 expended per new subscriber in 1994 may provide a rough estimate of the cost of building new cellular capacity. Personal communication from Cellular Telecommunicatons Industry Association.

Table 7-2. Value of Cellular Franchises Under Alternative Scenarios
Dollars

Annual cash flow per subscriber (before marketing)	*Investment per new subscriber*	*Marketing cost per new subscriber*	*Value per pop (@ 15 percent discount rate)*	*Value per current subscriber (@ 15 percent discount rate)*
550	700	600	251	5,020
311	700	600	70	1,400
205	700	100	45	900

All calculations assume annual:
Current subscribers per pop = 0.05
Subscriber growth rate = 0.17
Cash flow decline rate = 0.05
Investment cost decline rate = 0.05
Marketing cost decline rate = 0.05
Annual reinvestment rate = 0.05
Terminal value multiple of net cash flow = 12

Current cellular use rates in the United States are generally 40 to 50 cents per minute of prime time use plus a nominal fixed charge. The rates for low-use customers are generally about $25 per month plus use charges. The average monthly bill has fallen to $56.21 per month, suggesting about 80 minutes of peak use (at 40 cents per minute) and a flat monthly charge of $25 (table 7-1).

Cellular Franchise Values

At these rates, for most U.S. residential subscribers cellular service is obviously not competitive with wire-line telephony. But the transactions prices in recent acquisitions and mergers suggest that rates could fall substantially. The ratio of cellular subscribers to the general population is currently about 0.093. Recent acquisition prices have been about $250 per "pop" (population), or about $5,000 per current subscriber (table 7-2, row 1).[10] Given marketing and start-up costs, it is reasonable to expect that current systems with no more than a competitive prospective return on capital would sell for about $1,400 per subscriber.[11] Current system transaction prices are therefore more than three times the full reproduction cost of the franchise. Were cash flows to decline by 43

10. The value per subscriber is calculated by assuming that each of two systems attracts 0.05 of the population served.

11. The marketing cost per new subscriber is about $600; the cost of start-up might add another $100 per subscriber.

percent, systems would still sell at their reproduction costs, assuming the same investment requirements (table 7-2, row 2).

Cellular as Local Access?

At a capital charge of 20 percent,[12] the annualized cost of the cellular plant with current use patterns is $140 per subscriber, but it is falling. The incremental capital cost for new subscribers is $700 or less and could be as low as $400 per subscriber in five years, even without the conversion to digital. The portable transceiver costs about $100 per subscriber, or $20 per year, and is also declining. Operating costs of cellular systems currently average $155 per subscriber.[13]

Two other components of current cellular costs are extremely high, largely because of the novelty of cellular service and the strategic and regulatory uncertainties in delivering it. Marketing costs average about $600 per new subscriber, or about $120 per year (assuming no churn), and general and administrative (G&A) costs are about $140 per ongoing subscriber.[14] Regulated local-exchange carriers (LECs) currently spend less than $100 annually per subscriber on these two functions. We would expect cellular systems to reflect these lower costs once cellular becomes a bona fide alternative to wire-line service, rather than a specialized or novelty service.

The annual non-traffic-sensitive (NTS) costs of cellular service may be approximated by transceiver costs, G&A, and marketing. These currently total $280 per year, assuming an annualization of marketing costs at a 20 percent rate. Operating and plant costs, which vary with busy-hour use of the plant, average $295 annually per new subscriber for systems with an average of 1,000 subscribers per cell. Currently, these subscribers use 80 busy-hour minutes of service per month; therefore, the *average* traffic-sensitive cost of each minute is about 31 cents. Thus, a $23.33 monthly flat rate—

12. Mitchell uses a 15 percent capital charge for telephone plant, but cellular systems surely have a shorter useful life and are subject to greater risk of competition from new technologies.

13. These estimates are adapted from Dennis Liebowitz and others, *The Wireless Communications Industry*, Donaldson, Lufkin & Jenrette, Winter 1994–95.

14. Ibid.

which includes the cost of the receiver—plus 31 cents per minute of use would cover the cost of cellular service for the typical subscriber today. With LEC-equivalent marketing and G&A costs, the monthly rate (including transceiver rental) could be as little as $10 per month plus 31 cents per minute of busy-period use. With marketing costs of only $100 per new subscriber, cash flows could be reduced by more than 60 percent from current levels and still allow cellular franchises to be valued at the reproduction cost of assets (table 7-2). Technical progress and increased use should lower the incremental cost of a busy-hour minute substantially; as a result, it should also lower the price of use to substantially less than 31 cents per minute.

Currently, the average telephone line—business and residential—records about 14,000 minutes of local use per year for both incoming and outbound calls.[15] Park and Mitchell found that business use per line was about twice the magnitude of residential use in a small community that would be the type of market in which cellular service could potentially compete with wire-line service.[16] If business lines are about 30 percent of all switched lines,[17] this would suggest that the average residential local use is about 10,800 minutes per year. Half of this traffic is outbound calls; if the average call is 3.5 minutes long,[18] the average number of local residential calls per line is about 1,500 per year.

We assume that cellular service will most likely emerge as a substitute for wire-line access in areas with relatively low calling volume and relatively long loop lengths (that is, in smaller communities). Park and Mitchell found that the small community they studied generated only 950 calls per residential line per year in 1976, but telephone use has undoubtedly increased since then. We have assumed that residential lines in smaller communities cur-

15. This estimate is derived from FCC data on dial-equipment minutes (DEMs) for all companies, as reported to them by NEA. See *Monitoring Report*, May 1993, chapter 4.

16. Rolla Edward Park and Bridger M. Mitchell, *Optimal Peak-Load Pricing for Telephone Calls* (RAND Corporation, March 1987).

17. FCC, *Statistics of Communications Common Carriers*, 1993–94 edition.

18. The average local call appears to be about 2.25 minutes in duration, given data on local dial-equipment minutes (DEMs) and local calls reported by the FCC. (*Statistics of Communications Common Carriers*, 1992–93 edition, p. 157, and *Monitoring Report*, May 1994, p. 264.) Residential calls are likely to be much longer than business calls; hence, we use 3.5 minutes for the average residential call.

rently generate 1,400 calls per year, or somewhat less than for the entire country.

Park and Mitchell ascertained that flat-rate local service induced residences to concentrate 55 percent of their calls in the peak period (10 A.M.–5 P.M.), 38 percent in the adjacent shoulder periods (8–10 A.M.; 5–6 P.M.), and 7 percent in off-peak periods. We follow their approach in estimating the shift of traffic generated by use-sensitive pricing.[19] Park and Mitchell conclude that the welfare-maximizing set of three-period prices for calls within a single switching center was 9 cents per call for the peak period, 3 cents per call in shoulder periods, and 0 cents in the off-peak period, all measured in 1984 dollars. (In 1993 dollars, these rates would be 12.6 cents, 4.2 cents, and 0 cents, respectively.) For calls between local switching centers a few miles apart, the optimal 1984 rates were 13 cents, 7 cents, and 0 cents, respectively, or 18.2, 9.8, and 0 cents in 1993 dollars.

Cellular use is clearly more expensive to provide than wire-line use. If the marginal cost of busy-hour cellular use were three times the welfare-maximizing price of wire-line use and the monthly charge were $25 (including receiver rental), rates would be 37.8 cents, 12.6 cents, and 0 cents per call in the peak, shoulder, and off-peak periods, respectively. Assuming that calls average 3.5 minutes, this suggests a peak charge of about 10 cents per minute. In table 7-3, we show the implied calling pattern and the revenues per subscriber, assuming that only the outbound caller pays for this use and both the caller and recipient pay the rates cited here (if all calls are originated and terminated by cellular subscribers).

Under the most expensive assumption—that both caller and recipient pay the use charge—the average monthly charge for local telephone service is only slightly more than that required under Ramsey pricing for flat-rate access to the local wire-line plant (chapter 3). At these prices, calling volumes fall by about 20 percent relative to their levels under flat-rate pricing, but peak-period calling falls by 48 percent. Nevertheless, under these assumptions total minutes of use are still substantially higher than those recorded by

19. Specifically, we follow Park and Mitchell, *Optimal Peak-Load Pricing,* and assume that the own-price elasticity of peak and shoulder period use is −0.06 and that 50 percent of calls shift to other periods in proportion to their frequencies under flat rates.

Table 7-3. Cellular Usage and Average Monthly Bills Using Three-Period Pricing

Monthly use under two pricing scenarios

Period	*Flat-rate use (calls/year)*	*Use with rates of $0.378 and $0.126 per call (calls/year)*	*Annual call revenues at $0.378 and $0.126 per call ($)*
Peak	770	403	152
Shoulder	532	584	74
Off-peak	98	127	0
Total	1,400	1,114	226

Monthly bills under two billing assumptions

	Payer of calling charges (dollars per month)	
Revenue source	*Calling party*	*Calling and receiving parties*
Monthly service fee	25	25
Calling revenues	19	38
Total	44	63

Note: All calculations are based on Park and Mitchell's methodology, assuming an own-price elasticity of demand equal to −0.024 and that 50 percent of calls shift to other time periods in proportion to ex ante calling shares.

current cellular users. *Peak*-period minutes are about 1,400 minutes per year per subscriber (403 times 3.5); *total* current cellular use is about 1,000 minutes per year per subscriber. This additional peak-period use would increase the required investment in switching plant and cell sites, undoubtedly accelerating the crossover to digital technology.[20]

Current rates are clearly higher than those simulated in table 7-3. This is partly because current cellular rates are substantially above long-run incremental costs. In 1994, systems generated about $550 in cash flow per subscriber per year, before capital expenditures and marketing expenses.[21] Were revenues to fall by $240 per subscriber per year, the value of current systems per pop would still have been $70 (table 7-2, row 2), and the value per subscriber would be equal to reproduction costs. If, in addition,

20. An analysis of the prospective costs of a high-capacity digital cellular system appears in the next section.

21. Liebowitz and others, *The Wireless Communications Industry*, p. 14. Author's calculations.

marketing costs were to fall to $100 per subscriber, annual revenues could fall by another $67 per year. Revenues per subscriber of somewhat less than $550 per year would be sufficient to amortize current investments. These revenues could be raised by the charges assumed in table 7-3 but require far more use than cellular operators currently experience.

Such additional use would require more investment in cellular capacity. However, the cost of this additional capacity for analog systems might be very high, and spectrum might be a binding constraint on expansion of the subscriber base. Current cellular systems are simply not designed for the subscriber use rates of wire-line systems. Cellular systems generate use rates of about 0.026 erlangs per busy period whereas basic telephony systems are about 0.05 to 0.1 erlangs, depending on the density of the market.[22] To accommodate these higher use rates, cellular service will undoubtedly have to switch to digital technology.

One might object that it is impossible to examine cellular service as a *potential substitute* for basic wire-line telephony, because it currently serves as a complement to basic telephone service. In fact an examination of Organization for Economic Cooperation and Development (OECD) data suggests that cellular may be a substitute for regular telephone service for subscribers in some countries.[23] A logit specification of the penetration of cellular services across 21 OECD countries finds that the elasticity of penetration with respect to the fixed monthly wire-line telephone rate is +1.1 and the elasticity with respect to the fixed monthly cellular rate is −1.1, but there is no effect of *use* rates for either service.[24]

Digital Cellular Service

As the demand for mobile or cellular telephone service grows and technology improves, cellular systems will be forced to replace

22. An erlang is a measure of use employed by telephone engineers; 0.1 erlang is equal to 3.6 CCS (hundred call seconds) per busy hour.

23. Organization for Economic Cooperation and Development (OECD), Working Party on Telecommunications and Information Services Policies, *Mobile and PSTN Communications Services: Competition or Complementarity?*, June 1993.

24. Authors' estimates from OECD data.

their current analog systems with digital technology. Reed has estimated that the prospective costs of building a digital cellular system with a market penetration of 10 percent is about $600 (in 1992 dollars) per subscriber.[25] This system would deliver 0.03 erlangs of service during busy periods. Reed's model suggests that increasing use from 0.03 to 0.1 erlangs at busy hours to approximate typical residential subscriber use would increase this cost to about $950 per subscriber. These prospective costs are based on foreseeable technological progress in the 1997–2002 period.

The annual operating costs of Reed's cellular system at 10 percent penetration is $400, including connections to the switched network at $0.03 per minute. A simple calculation thus suggests that cellular systems will soon offer wire-line-grade service for less than $50 per month.[26] This is very close to the Ramsey quasi-optimal rates for current telephone companies in areas at the 75th percentile for loop length (table 7-4).

A recent study prepared for AT&T, MCI, and CompTel came to vastly different conclusions concerning the capital cost of cellular: cellular telephony service would require an investment of $2,800 per subscriber (or more than three times the current cost) to deliver local service at current wire-line use rates.[27] The principal contributor to this cost is enormous cell-site costs of $2,160 per subscriber, apparently based on the assumption of continuing use of analog technology. If this estimate is correct, it certainly explains why Reed assumes that future cellular service will be delivered over digital systems.

25. David P. Reed, "Putting It All Together: The Cost Structure of Personal Communications Services," Working Paper 28 (FCC Office of Plans and Policy, November 1992), p. 37.

26. Reed's estimates may be conservative. Cybertel, a cellular company in Kauai, Hawaii, is proposing to offer a digital service at $0.036 to $0.045 per minute. See In the Matter of the Application of CYBERTEL CORPORATION, dba CYBERTEL CELLULAR, *Application and Certificate of Service*, Before the Public Utilities Commission of the State of Hawaii, Transmittal No. 94-01, June 24, 1994.

27. Economics and Technology, Inc., and Hatfield Associates, *The Enduring Local Bottleneck.*

Table 7-4. Prospective Economics of Alternative Local-Loop Technologies for Urban Markets[a]
1993 dollars

Cost per subscriber	*Current telephone companies*	*Rebuilt telephone companies*[b]	*PCS*[c]	*Wireless fixed-access*	*Digital cellular service*[c]	*Cable telephony*[d]
Investment	1,100	910	900	800	950	565
Annual operating costs	360	360	400	420	400	420
Annual revenue requirement	580	540	580	580	590	533

Sources: David P. Reed, *Residential Fiber Optic Networks: An Engineering and Economic Analysis* (Boston: Artech House, 1992); David P. Reed, "Putting It All Together: The Cost Structure of Personal Communications Services," Working Paper #28 (FCC Office of Plans and Policy, November 1992); Dan Margiotta, "Wireless Fixed Loop Access: The Bridge to the Information Age," *Telephony* (August 2, 1993); George Calhoun, *Wireless Access and The Local Telephone Network* (Boston: Artech House, 1992); Economics and Technology, Inc., and Hatfield Associates, *The Enduring Local Bottleneck* (Boston, 1994).

a. All calculations assume a 20 percent annualization factor for capital investment.
b. Fiber optics and paired copper wires.
c. Assumes 10 percent market penetration; peak-hour usage equivalent to that of wire-line systems.
d. Assumes 100 percent penetration, using EEI/Hatfield capital costs.

Fixed Wireless Systems

Cellular telephony developed strictly as a *mobile* service, designed as a complement to traditional wire-line service. Its analog technology is far from ideal for delivering reliable telephony service. The wire-line network is rapidly evolving into a digital network for all but the subscriber loop, and wireless telephony will also likely be based on digital technology in the near future.

It is feasible to design a wireless technology for delivering telephone service to fixed and mobile subscribers; indeed, wireless systems are being built to replace decaying copper-wire loops and to extend service into rural areas. Because copper-wire loops increase in cost with loop length, the cost of providing service to rural areas can be prohibitive. Wireless systems, by contrast, radiate radio signals over a wide area, and their costs depend on the number of users per unit of spectrum, not on how far those users are from the central office.

Basic Exchange Telecommunications Radio Service

In the 1980s, rural telephone companies began to consider using more efficient radio-based technologies to extend or replace their plant. The Rural Radio Task Force was formed in 1985 to make a recommendation to the FCC for using spectrum for rural telephony. Its final report in June 1986 provided the underpinning for an FCC rule making that culminated in 1988 with the creation of a new class of service—basic exchange telecommunications radio service (BETRS). Spectrum in the 150 MHz, 450 MHz, and 800 MHz bands was designated for BETRS use, but thus far only the 450 MHz band has been used.

The first two BETRS systems were installed in 1986, two years before the final FCC order. Both were extremely small and in rural areas—Glendo, Wyoming, and Allen, Kansas. A number of small telephone companies have since installed these digital wireless systems to provide service to a relatively small number of subscribers.[28] In addition, Bell Atlantic is using a similar technology to wire rural areas of Mexico.

28. See Calhoun, *Wireless Access*, table 9-2a.

The economics of BETRS are still somewhat untested. Calhoun estimates that wireless access now costs about $3,000 per subscriber for normal telephone blocking rates and site costs.[29] Copper-wire plant costs are typically about $1,200 to $2,000 per loop and can be more than $5,000 per loop in rural areas. Wireless systems have numerous other advantages. First, they can be more readily adapted to deliver higher-speed data services, such as basic rate integrated services digital network (ISDN). Second, they do not require construction of expensive feeder and distribution systems that are not easily retrieved when communities decline. Third, they do not require large expenditures on feeder and distribution systems in anticipation of growth. Finally, maintenance costs are much lower for widely spaced antennas than for miles of aerial cable plant.

Currently, BETRS is competitive with wire-line only in areas of relatively low population density. However, given that digital radio-based systems enjoy a much more rapid rate of technical progress than do wire-line systems, this wireless technology will likely spread to more densely populated areas in the next few years. Calhoun expects BETRS systems to cost about $2,000 per subscriber within five years, while the cost of wire-line systems continues to rise with inflation. By 1997, BETRS could be competitive with copper-wire systems for rural telephone subscribers, but not for combined broadband/voice services now being developed in urbanized areas.

Urban Wireless Systems

A slightly different wireless system has been designed by NYNEX for urban areas.[30] This system is clearly aimed at reducing the use of copper wires and the attendant maintenance costs in high-cost areas. The system would use a fiber-optics backbone to connect a series of radio ports mounted on telephone poles or other facilities that can communicate with remote subscriber units at the subscriber's residence or business. NYNEX's initial trials with this system in a densely populated area of Brooklyn, New York, have been

29. Ibid.
30. Margiotta, "Wireless Fixed Loop Access," pp. 20–26.

successful, and an associated economic study places the prospective costs of this technology at about 12 percent less than current copper-wire technology. But it is unlikely that this technology would be developed by a firm *competitive* with a local wire-line telephone company. It involves a large amount of sunk costs and simply replicates traditional fixed telephone service and offers nothing new in terms of portability or service attributes, such as broadband applications.

Personal Communications Services

Other radio-based technologies for providing access are now under development in the United States and Europe. These technologies are generally described as personal communications services (PCS) because the *subscriber* is addressed rather than a specific location, in much the same fashion as with portable cellular telephony.

PCS systems may develop in a variety of ways. One of the first commercial applications of PCS is Telepoint, a portable telephone service offered in the United Kingdom. A subscriber to Telepoint carries a receiver and accesses the network through a large number of dispersed low-power transmitters in major buildings or public facilities. This service is essentially a cellular service with extremely small cells.

Other applications may be developed for simply extending the reach of current cordless telephones or connecting to a satellite-based interexchange facility. Cable companies have shown interest in PCS as a means of developing voice telephony services over their cable plant. Established telephone companies apparently want to use PCS to extend their basic service through more portable handsets.

The FCC recently finalized its rules for allocating spectrum to PCS in the United States. This process has moved much faster than the cellular licensing phenomenon of the 1970s and 1980s, because the federal government wants to raise revenues from new spectrum allocations in the name of deficit reduction. The Omnibus Budget Reconciliation Act of 1993 required the FCC to auction the PCS spectrum within 210 days of the act's passage. The goal was to raise an estimated $11 billion through this FCC auction. Federal budget deficits have thus finally created an incentive for the U.S. government to rely on competitive markets rather than regulation to allocate spectrum.

The auctions of spectrum for PCS will involve three 30 MHz blocks and three 10 MHz blocks. The auction of two of these 30 MHz blocks for each of 51 metropolitan trading areas (MTAs) has now been completed, raising $7.7 billion.[31] It now appears that this spectrum will be used primarily to allow existing cellular companies to develop national service offerings. For instance, AT&T (the nation's largest cellular carrier) had the winning bid for one of the two licenses in nineteen of the fifty-one MTAs. Wireless Company—a consortium of Sprint (the third largest U.S. long-distance company) and three cable television companies—was the winner of twenty-nine blocks covering nearly three-fourths of the United States; a consortium of four cellular companies, PCS PrimeCo,[32] was the winner of eleven blocks.

Reed estimates that stand-alone PCS systems will cost slightly less than the digital cellular system described previously, in part because they do not generally require high-speed handoff between cell sites for vehicular use. Reed estimates that capital investment will be about $700 (in 1992 dollars) per subscriber for 0.03 erlang busy-hour use and $900 per subscriber for 0.1 erlang service.[33] The ETI/Hatfield study places the initial cost per subscriber at $1,100, largely because of higher assumed customer receiver costs.[34] Operating and annualized capital costs are somewhat less than $600 (table 7-4).

At these costs, PCS is potentially a major source of contestability for local access in voice and low-speed data applications. It is not likely to compete with the integrated fiber/coaxial broadband networks being built by cable and telephone companies, but it could

31. This includes three blocks allocated under the controversial "pioneer's preference" program. The average price paid per pop for the first two 30 MHz blocks was $15. The auction of the third 30 MHz block has been delayed because of a legal challenge of the FCC's minority and small-business preference policy that was to govern this auction.

32. The four companies are Bell Atlantic, NYNEX, U S West, and Air Touch (formerly owned by Pacific Telesis).

33. Reed, *Putting It All Together*, pp. 21–24.

34. Economics and Technology, Inc., and Hatfield Associates, *The Enduring Local Bottleneck*, p. 89.

Table 7-5. Forecasts of Growth of Wireless Local Access in the United States, 1992–2003

Thousands

	1991 forecast Cellular subscribers at end of year							
Year	*1993*	*1994*	*1995*	*1996*	*1997*	*1998*	*1999*	*2000*
Subscribers	13,350	16,850	20,400	24,050	27,800	31,600	35,400	39,200

	1995 forecast Wireless subscribers at end of year										
Year	*1993*[a]	*1994*	*1995*	*1996*	*1997*	*1998*	*1999*	*2000*	*2001*	*2002*	*2003*
Cellular subscribers	16,009	23,600	31,485	38,910	46,110	52,170	57,820	63,140	68,130	72,900	77,670
SMR subscribers	0	15	115	915	2,115	3,615	5,115	6,615	8,115	9,615	11,115
PCS subscribers	0	0	15	290	1,890	4,830	8,680	12,860	17,370	22,100	26,830
Total wireless subscribers	16,009	23,615	31,615	40,115	50,115	60,615	71,615	82,615	93,615	104,615	115,615

Source: Dennis Liebowitz and others, *The Cellular Communications Industry* (Spring 1991), p. 80; *The Wireless Communications Industry* (Winter 1994–95), p. 13, Donaldson, Lufking, and Jenrette.

a. Actual subscribers.

discipline the telephone companies that attempt to raise residential access rates for voice services. Because it is a wireless technology, very little of its capital needs to be sunk before service demand materializes. Most of the $900 (or, at most, $1,100) capital cost for 0.1 erlang service is in the customer connection and the switching capability in the cells. The outlays on both are related directly to the number of customers subscribing to the service.

The rapid development of wireless technologies and their declining costs provide opportunities for developing a truly contestable local telephone industry. Forecasts of the growth of the wireless telephone have proved far too conservative. (See table 7-5 for an example). By the end of the century, there will likely be one wireless telephone for every three Americans and Canadians.

Cable Television

Cable television is almost ubiquitous in the United States and Canada. More than 90 percent of all households are passed by cable television, and nearly two-thirds subscribe. These cable systems are designed to provide one-way video delivery and (in some cases) a limited amount of two-way communications. However, none of these systems has been designed to provide switched (telephony) services.

Recently, cable operators have begun to look for new technologies that would allow them to offer a more complete array of telecommunications services. Given the large capacity of their subscriber drop lines (which are generally coaxial cables) and their backbone distribution plant (which is rapidly being converted to fiber optics), these cable systems are in a position to develop a wide array of new services. Telephone companies have recognized these opportunities and are now investing in cable systems outside their franchise areas (table 7-6) to exploit their expertise in switched services.

At present, the only examples of cable television switched-access telephone services are found in the United Kingdom. U S West, Southwestern Bell, Comcast, Bell Canada, and NYNEX have invested in two new U.K. cable television companies and are offering telephone service as an option to regular cable television service. There are projections that these cable television companies will

Table 7-6. Telephone Company Acquisitions of Major Cable Television Firms

Acquiring telephone company	*Cable television acquisition*
Bell Atlantic	TCI/Liberty Media (abandoned)
U S West	Time Warner (25%)
Bell Canada Enterprises	Jones Intercable (30%)
SBC (Southwestern Bell)	Hauser Communications
SBC	Cox Cable (40%)
Bell South	Prime Cable (22.5%)

attract as many as 600,000 telephone subscribers in the United Kingdom by the end of the decade.

The estimated costs of building two-way switched telephony services into an existing cable system range from $300 to $745 per subscriber.[35] These costs are the *incremental* expenses of adding telephony to a fiber/coaxial network, including telephone-switching capacity, a network interface unit at the cable system's headend, and a customer interface unit that directs the cable system's signals to the various household television and telephone receivers. All of this equipment (except the telephone switch) is currently in the development phase; the cost of all of it is likely to drop over time.

For a new cable system, such as those being built in the United Kingdom, the cost of adding switched-access telephone services is estimated at about 22 percent of the cable plant—perhaps $300 to $350 in U.S. dollars. Reed's estimate of the incremental cost of a PCS system built over an existing cable network is about $450 per subscriber.[36] This cost is clearly de minimis compared with all other alternatives for providing access to the telephone network. With 100 percent penetration, even the highest cost estimate ($745) falls to $565 per subscriber. Perhaps the largest stumbling block to

35. The lower estimate is based on trade-press reports, such as "Will the Broadband Network Ring Your Phone?" *Telephony*, December 6, 1993, p. 34, and pronouncements by Scientific Atlanta, an equipment supplier. The higher estimate is found in Economics and Technology, Inc., and Hatfield Associates, *The Enduring Local Bottleneck*, table 3.3.

36. Reed, *Putting It All Together*, p. 35. A more recent estimate of $400 to $500 per subscriber was presented by Johanne Lemay of Lemay-Yates Associates ("Evaluating New Technologies and Their Commercial Consequences") at the May 1994 Canadian Telecommunications Superconference, "Towards a New Era of Telecompetitiveness" (Toronto, Ontario, The Canadian Institute).

cable-company telephony is the difficulty cable systems may have in marketing the service. The ETI/Hatfield analysis concludes that such a service will probably not be profitable, because of steep marketing costs as a result of low subscriber demand for a cable-provided telephone service. It assumes that a cable company, after fifteen years of marketing, will only attract 18 to 30 percent of the market it serves, at a present-value cost of $183 to $301 per eventual subscriber (at 14 percent discount rates). Were a cable company to attract subscribers at a marketing cost of just $200 per *additional* current subscriber, such a system would be enormously profitable even with only 18 percent penetration.

Because we cannot know the costs of marketing a new cable telephony service, we assume in table 7-4 that annual operating costs (including marketing) are at the high end of the other services' costs—$420 per year. In addition, we use the full market penetration cost of $565 per subscriber to compare it with the ubiquitous current telephone company service. Under these assumptions, cable telephony has a lower revenue requirement than even the most efficient, rebuilt wire-line telephone network.

Satellite Technologies

Yet another way to use the electromagnetic spectrum to provide ubiquitous access to the telephone network would be through satellites, which could be accessed directly by portable handsets or by fixed antennas. Many large users have very small aperture terminals (VSATs) to connect hundreds or even thousands of remote sites such as hotels, motels, or other commercial enterprises for data transmission, including credit card verification, reservations, or inventory information. In addition, new national paging services (such as Skytel) are being built around satellite links.

Using satellites to provide universal access to the network for voice services is hampered by the delays in transmitting a signal to a satellite transponder in geostationary orbit and back. An alternative is to use a number of low-orbit satellites, a strategy that Motorola is pursuing through its IRIDIUM Project. However, we have no data on the likely cost of this approach to wireless access.

Table 7-7. Competitive-Access Providers (CAPs) in the United States

Company name	*Ownership*	*Cities served*
Bay Area Teleport	Pacific Telecom	San Francisco
City Signal	Various	Grand Rapids, Detroit
Diginet	Harold Sampson	North Shore of Chicago, Milwaukee, Kenosha, Racine
Digital Direct	TCI	Seattle, Dallas, Chicago, and Sacramento
Eastern Telelogic	New York Life, others	Philadelphia and suburbs
Electric Lightwave	Citizens Utilities	Portland, Seattle, Salt Lake City (future), Sacramento (future)
Fiber Optics Corp. of the U.S.	Not available	Philadelphia
Fibernet, Inc.	Time Warner	Rochester, Buffalo, Albany, Syracuse
Hyperion Telecommunications	Adelphia Cable, Continental Cablevision	Jacksonville
Institutional Communications	MFS	Washington, D.C.
Indiana Digital Access	Time Warner, others	Indianapolis, Terre Haute
Inter-Media Communications	New York Life, others	Orlando, Tampa, Miami, Jacksonville
IOR Telecom	Iowa Resources	Des Moines
Jones Lightwave	Jones International	Atlanta, Chicago, Tampa
Kansas City Fibernet	American Cablevision	Kansas City, Independence, Mo.
Metrex Corp. of Atlanta	MFS	Atlanta
Metrex	Various	Birmingham, Huntsville

Competitive-Access Providers

One of the most pervasive and important of all the alternative access media is the fiber-optics network in most major U.S. cities. These networks have been developed to provide interconnection of large local area networks, one-way data transport, and voice-data service access to interexchange carriers. Their customers are large businesses that generate substantial amounts of traffic. However, the networks are not an alternative for dispersed small business and residential customers; it is unlikely that any company would build such systems to connect users who generate such small amounts of traffic. Because they are essentially unregulated carriers, few public data about them exist. Huber, Kellogg, and

Table 7-7. (contd)

Company name	*Ownership*	*Cities served*
Metro Com	n.a.	Columbus, Cleveland, Akron
Metropolitan Fiber Systems	Peter Kiewit Sons, Inc., various	Atlanta, Baltimore, Boston, Cambridge, Chicago, Crystal City, Va., Dallas, Houston, Los Angeles, Minneapolis, New Carrollton, Md., New York, Pittsburgh, Philadelphia, Reston, Va., San Francisco, Washington, D.C.
New England Digital Distribution	n.a.	Boston, Cambridge
Ohio Linx	Various	Cleveland, Akron, Dayton, Toledo
Penn Access Corp.	TCI	Pittsburgh
Phoenix Fiberlink	Phoenix American	Sacramento
Teleport Group	Cox, TCI	Baltimore, Boston, Cambridge, Chicago, Dallas, Garden City, N.Y., Houston, Jersey City, Los Angeles, Newark, New York, North Brunswick, Princeton, San Francisco, Weehauken
Teleport Denver	Intertel Communications	Denver
Western Union ATS, Inc.	MCI	Chicago

n.a. Not available.

Thorne provide some fragmentary data from private commercial studies (table 7-7).

The potential cost of equipping CAP fiber rings for ordinary telephony has been estimated by ETI/Hatfield Associates at $1,210 (in 1994 dollars) per subscriber, more than half for backhaul (two-way distribution) facilities.[37] This estimate may be high; but because CAPs are not an alternative for the dispersed residential and small-business customer located away from the city center, such estimates are of minor importance.

37. Economics and Technology, Inc., and Hatfield Associates, *The Enduring Local Bottleneck*, p. 89.

Can Access Markets Be Truly Contestable?

For a market to be contestable, an incumbent firm must be constrained from exercising any monopoly power because of the prospective effects of new entry into the market. This new entry need not be sufficient to wrest the *entire* market from the incumbent—merely enough to render unprofitable any increase in price above the competitive (long-run incremental cost) level.

For local telephony, contestability does not require that any one potential entrant be able to seize a large share of the local telephone company's market if the latter raises rates. Rather, an entrant (or several potential entrants) must be able to discipline any rate increase to a given set of customers, thereby rendering such an increase unprofitable and unlikely. Thus, for example, CAPs may keep carriers from attempting to raise rates on high-speed (that is, T-1 or greater) lines to large business customers. Alternatively, new PCS systems may provide potential rate discipline for low-volume urban residences, or the threat of cable television entry may serve a similar function for higher-volume users.

Were rates deregulated, the incumbent telephone carriers would be forced to assess any prospective rate increase by weighing the loss of net revenues from lost subscribers against the increase in gross revenues from remaining subscribers. If incremental costs of service are low, a prospective 10 percent rate increase for access service might prove unprofitable in the short run if 11 or 12 percent of subscribers were bid away by competing local-access providers. The decision is made even more complicated because the incumbent must weigh any beneficial short-term effects of a rate increase against the loss of business *over time* and on the probability of triggering rate reregulation.[38]

An Analysis of Prospective Entry

A glance at table 7-4 suggests that incumbent telephone companies will not only witness substantial market entry in the next

38. The 1992 Cable Act should serve as a reminder to any deregulated firm that deregulation is not a permanent state. In 1984, cable television was deregulated by a statute passed by Congress and signed by Ronald Reagan. In 1992, the industry was reregulated by Congress over the veto of George Bush.

few years, they are already in some difficulty. Their current plant investment averages $1,100 per line, but new technology could reduce this to $910 per line or even less.[39] Regulation is thus attempting to amortize an embedded plant whose replacement cost is falling over time—clearly a risky proposition in an era of increasing competition.

Assuming an annual capital charge of 20 percent to amortize the investment of new, competitive technologies, the revenue requirements of the various wireless technologies listed in Table 7-4 appear to be below that of current telephone companies and somewhat below the $620 in average revenues per line for current telephone companies.[40] There is thus mounting evidence that wireless technologies will soon be competitive with the current wire-line services. Cable television offers prospectively lower costs than even the "rebuilt" wire-line telephone systems.

Table 7-4 also suggests that new wire-line telephone company plant, employing a combination of fiber optics and copper wires, will be nearly 20 percent cheaper than the value of the current embedded plant.[41] But telephone companies will probably not want to replace their plant with a new set of facilities to deliver only the current set of telephone services. However, they may wish to replace it with a plant capable of delivering a *wider* array of services, such as video. Indeed, most major companies are contemplating precisely such a move.

Will competitive wireless services attempt to compete with the telephone companies in this environment? Given that the wireless services cannot deliver broadband services with current technology, they must consider the likely response of the telephone companies to such entry into the market. As the telephone companies replace their narrowband networks with a plant capable of deliv-

39. Reed, *Residential Fiber Optic Networks*, appendix tables B.2 and B.3.

40. Some of this $620 derives from third-party billing services, directory advertising, and miscellaneous services not directly associated with local access-exchange service.

41. Even this estimate of the replacement cost of current plant may be too high. Pacific Telesis is proposing to build a new fiber/coaxial cable system, capable of delivering video services as well as standard telephony, for less than $1,000 per subscriber. See FCC, *In re Application of Pacific Bell* (1993); and Robert G. Harris, Testimony in Support of Pacific Bell's Section 214 Application to the Federal Communications Commission, December 1993.

ering video services, they may be in no position to fend off entry in narrowband telephony services through deep rate cuts. Nor will they be able to raise traditional telephone rates far if the prospective costs of wireless services are in the range shown in table 7-4. The telephone plant is largely sunk; the estimated investment costs for wireless include a much larger share of costs that are not sunk.[42]

Nor can the probability of entry be estimated simply by looking at prospective *average* costs. Surely wireless technologies are already much cheaper for connecting low-use subscribers in rural areas. Fiber-optics rings (CAPs) are growing rapidly in dense commercial areas with a heavy concentration of voice and data traffic; digital service is already moving into some areas where demand is most intense. In short, no single technology is optimal for all access services. If local telephone companies are deregulated,[43] the incumbent local telephone companies might or might not find it profitable to raise access rates substantially above average incremental costs for any identifiable group of subscribers, given the wide variance in use patterns.

Ultimately, development of competitive access may eliminate most forms of government regulation of the local loop. But even though many new technologies for providing access are now falling in cost to a level that might be attractive relative to telephone-company wire-line technology, they may not be developed by new entrants for two related reasons. Current access rates for residential subscribers are below the average incremental cost of wire-line service, reducing the attractiveness of entry. The established telephone companies' costs are also largely sunk, allowing them to price aggressively (regulators permitting) if market entry occurs.

In principle, the first of these problems is easily remedied. Regulators can recognize that they are a major source of entry barriers and force a repricing of service despite the immediate political unpopularity of such a move. It may be naive to think that they will do so until the sources of revenues for the cross subsidy erode. But there are hopeful indications that states are beginning to recognize that continuing these cross subsidies is impractical and unwise.

42. For instance, about 20 percent of the cost is the portable handset, which obviously can be recouped through a sale in other jurisdictions.

43. In chapter 8, we point out that such deregulation will probably have to include mandatory interconnection.

Current federal legislation in the United States, requiring further opening of intrastate toll markets to competition, may accelerate this trend.

The second problem is more formidable. The experience in cable television markets suggests that entrepreneurs will not build an expensive wire-line network next to an established system—the sunk cost problem looms too large. Of course, the growth in CAPs provides a counterexample; surely the telephone companies can and do reduce prices to meet competition from these fiber-optics networks. It is also possible that new radio-based systems, with less sunk costs, might present more attractive investment opportunities. In addition, adding switched telephony to an existing cable plant may be quite attractive. Many telephone companies are eagerly contesting for the opportunity to show cable operators how to do just that—outside their own franchise areas, of course.

For the typical resident in an urban area, cable television companies would appear to be the principal source of competition to local telephone companies. However, the ETI/Hatfield study concludes that telephony will not be an attractive investment for cable operators. Their conclusion follows from an assumption that these cable operators will not control more than 10 percent of the telephone-access market, and even this small share only after a considerable period of time. In the United Kingdom, by way of contrast, cable companies are attracting 20 percent of their subscribers to their telephone service even though British Telecom's local-access rates cannot go up. Were British Telecom to raise local rates significantly, there would surely be a larger shift of telephony to cable operators.

How much of a shift of patronage would have to occur to defeat a telephone rate hike following deregulation? This depends on what share of the incumbent telephone company's costs are fixed and, perhaps more importantly, what loss in other revenues would follow from the loss of a single telephone company subscriber. Even assuming that the market elasticity of demand for access is zero, a telephone company would not raise rates by 10 percent if such an increase led to a loss of at least 13 percent of subscribers to rivals if the telephone company's variable costs were one-third of revenues. But if current or prospective cash flows from services other than local flat-rate monthly service were half of total cash flows from each subscriber, a 6.5 percent loss of subscribers would be enough to defeat the rate increase. Given the potential of new services such as video-on-demand, these nonaccess revenues may begin to loom large in the decisionmaking of even an unregulated local-access provider. Hence, even if local telephone companies continue to have a large share of ordinary telephone connections, they may be wary of increasing local rates for fear of losing a variety of revenue sources to competing media.

Conclusion

Huber, Kellogg, and Thorne have suggested that the U.S. AT&T decree is based on the wrong facts: monopoly in local access and competition in interexchange services.[44] They argue that the local loop is now essentially contestable but that AT&T would reestablish its interexchange monopoly, due to the economies of scale in fiber-optics transmission. Although this position may be a little extreme, it certainly reflects the trend in delivering local access. Though most subscribers continue to face but one reasonably priced access alternative in 1995, these options will surely increase in the next few years. In the United States, consumers will soon be able to choose among as many as six different wireless providers—two cellular, one SMR, and three PCS services. Improvements in digital spectrum-based and fiber/coaxial-cable technologies are threatening to burst the access bottleneck. Were local access priced at incremental cost or at Ramsey quasi-optimal levels, this competition would come even sooner.

44. Peter W. Huber, Michael K. Kellogg, and John Thorne, *The Geodesic Network II: 1993 Report on Competition in the Telephone Industry* (Washington: The Geodesic Company, 1992), p. 6.45.

Chapter Eight

Regulating Telephony in the Era of the Information Superhighway

---

IN THIS BOOK we have examined U.S. and Canadian regulatory policies toward voice telephony services delivered over the traditional network through twisted copper-pair distribution lines. But even the most casual observer of telecommunications recognizes that telecommunications networks are rapidly evolving into a much more complex set of institutions, capable of delivering voice, data, and video services through wires, coaxial cables, optical fibers, and the electromagnetic spectrum. Telephone companies are faced with increasing competition in their core voice-data businesses and an exploding array of new opportunities for new technologies and services. But these opportunities create substantial regulatory difficulties (outlined in chapter 4), as competitors complain that they are being exploited through cross subsidies from the older, traditional regulated services.

In the first three chapters of this book we described the regulatory policies in Canada and the United States and documented their large social costs. For decades, consumers were denied the benefits of competition. They faced telephone rates contrived by regulators to minimize political opposition, at a cost of billions of dollars of reduced economic welfare. To sustain these policies, regulators were also forced to block entry of new carriers and new technologies in this dynamic sector. The United States came to its senses first, allowing a limited amount of competition in the 1960s and 1970s. But a major antitrust suit was essentially responsible for today's competitive structure in terminal equipment and long-distance markets. Eventually, the U.S. Federal Communications Commission recognized that the entire telephone rate structure had been distorted by regulators, and in 1984 it began a modest rebalancing of rates to correct decades of bad policy.

Canada is following the U.S. lead, albeit with a lag of at least a decade. However, it is still far from clear that Canadian regulators recognize how far they still must travel. Rates were and are much more distorted in Canada than they ever were in the United States. Private-network competition is impeded by a much more restrictive spectrum-allocation policy. Provincial governments put a harder brake on progress than did state commissions in the United States, which are finally beginning to address the shortcomings of their restrictive policies. The newly defined federal jurisdiction and the new Telecommunications Act suggest a faster evolution toward competition in Canada than would have been the case under the old divided regulatory regime. However, the Canadian government intervened to delay for one year the first bold venture by the Canadian Radio and Telecommunications Commission (CRTC) into a more rational pricing of local and long-distance service.

Given this sorry record of regulation—repeated many times over in the transportation, finance, and energy sectors—it is difficult to be optimistic about current government efforts to regulate the telecommunications industry in the name of a level playing field. Cable television companies, cellular carriers, incumbent local telephone companies, new and old long-distance carriers, satellite companies, equipment suppliers, and even software suppliers[1] are bidding to participate in the exciting new telecommunications marketplace. But government restrictions on telephone-cable competition, on vertical integration by telephone companies, and (in the United States) on the geographic scope of many major cellular companies create enormous barriers to developing new alliances to enter these markets. Each of these restrictions is justified as a prophylactic device to prevent monopoly exploitation, but each reduces the rate of entry and, potentially, impedes technical progress and competition.

In this chapter, we describe some of the new technologies and markets toward which telephone companies are gravitating and the regulatory problems they create. We rely on the theoretical discussion of chapter 4 to guide policy prescriptions, most of our discussion relating to the prognosis for the United States.

1. Such as motion picture companies, publishers, and cable television networks.

What Is the Information Superhighway?

It is fashionable to think of all communications as evolving toward some extremely fast, sophisticated system that would provide all users with the opportunities to interact at rates of many megabits per second (Mbs) or even gigabits per second (Gbs).[2] In thinking about this high-tech future, however, it is useful to remember that most current communications take place at only a few kilobits per second (Kbs). Indeed, even the highly touted integrated services digital network (ISDN) is generally offered at only 128 Kbs (two channels of 64 Kbs, plus a control channel) and used by only a small fraction of business firms. More common are the 14.4 Kbs modems connected with remote computers or commercial databases such as Prodigy or CompuServe.

Transmission Speed

Modern technology allows telecommunications carriers to transmit and even switch information at very high speeds. Modern digital transmission systems offer speeds of DS-3 (45 Mbs) and even higher. European, Japanese, and U.S. companies are developing switched services with even higher rates up to 1 Gbs. In the United States, carriers are preparing to install synchronous optical network (SONET) transmission systems that deliver digital signals at 51.84 Mbs to 2.488 Gbs.[3]

These high speeds are made possible by innovations in electronics and fiber optics that can be built into the backbone transmission and switching systems of modern telecommunications at an affordable cost. However, building these speeds into every feeder and distribution loop in the network could be quite expensive. As the copper-wire system degrades, carriers are replacing more and more of it with fiber optics, but few are installing fiber optics to all final consumers. The most aggressive plans (such as that of Pacific

2. Megabits are millions of units (bits—that is, 0,1) of information. Gigabits are billions of bits.

3. John Rivenburgh, "Paving the Way for Phase 2 SONET Deployment," *Telephony* (June 14, 1993), pp. 40–46.

Bell, which is detailed below) would combine coaxial-cable drop lines with fiber optics to provide ubiquitous broadband networks.

A less aggressive strategy that may prove feasible involves upgrading the existing paired copper-wire loop. In recent years, Bellcore and others have developed a technique for forcing video-quality signals through the traditional copper-wire subscriber loop. This technology, asynchronous digital subscriber loop (ADSL), is still in the developmental stage. Eventually it may give established telephone carriers the opportunity of using their current loops to provide video service, perhaps as leased-access service or video-on-demand.[4]

Development of more sophisticated, high-speed communications systems continues. The proliferation of local area networks (LANs), connecting computers, workstations, and databases within a single establishment or firm, has created the demand for high-speed interconnections. These LANs use widely different technologies but must be interconnected for a variety of business or scientific purposes. These connections may be made simply by dedicated unswitched private lines at varying speeds (depending on the customer's use) or through the switched network. In addition, many telephone companies are installing high-speed asynchronous transfer mode (ATM) switches for transmission and switching of discrete data packets. This service permits more efficient use of transmission and switching capacity for information transmitted in discrete noncontinuous bursts.

The Internet, a loose interconnected group of computer networks developed by a consortium of public and private entities under the direction of the National Science Foundation, has been one of the most successful innovations leading to the information superhighway. This network of networks has grown without a definite plan, allowing users to gain access to a large number of remote databases, transmit electronic mail, transfer files, and even engage in on-line conferences. Although the Internet can be accessed through commercial services at the low bit rates obtainable through regular telephone service, most users are connected

4. This service has been approved on an experimental basis by the Federal Communications Commission. It would allow telephone companies to offer video circuits for third-party lease, but they would not be permitted to participate in programming these channels.

through LAN gateways operating at much higher speeds, though still lower than those envisioned for the information autobahn.

Services

The new high-speed networks have obvious applications for scientific and commercial customers. But are there broad consumer uses that would justify building out these networks to a large number of residential lines?

The obvious consumer applications involve video. Conventional cable television services, which transmit scores or even hundreds of channels of entertainment, news, sports, and other information simultaneously, could be offered over a ubiquitous fiber-optics system or a mixed fiber/coaxial-cable system. Alternatively, the telephone network could be configured to offer video-on-demand through an ADSL local-loop technology and many video servers located at the central office. The latter form of video would essentially be a greatly expanded version of current pay-per-view systems. Telephone companies could provide either system directly or through leased access to a number of different video-service companies. However, there is some doubt about whether a telephone company would be willing to invest the capital to expand its telephone systems in either fashion, given the advantages of existing cable television companies.[5]

Of course, telephone companies could develop new consumer services that exploit the greater capacity of the information superhighway, but no one is sure what these services are. Videotelephony or videoconferencing would probably be of limited value to most residential customers. The transmission of vast amounts of data in a very short time is clearly useful for clients solving complex climate models or conducting remote medical diagnosis using magnetic resonance imaging (MRI), but these are not residential services. Perhaps individual customers will want to play complex video games or puzzles across town or across the country, but such novel uses would hardly seem sufficient to drive an investment of several hundred dollars or even more per residential access line.

5. For a discussion of this issue, see Leland L. Johnson, *Toward Competition in Cable Television* (MIT Press and American Enterprise Institute, 1994), pp. 87–110.

Given the inability of the Bell operating companies (BOCs) to exploit their newly won freedom to provide information services over their own networks,[6] there is reason for skepticism about the ability of these companies to launch expensive new broadband services. Yet other more entrepreneurial companies, such as motion picture or cable firms, could contract for channel access in advance or even form joint ventures with the telephone companies, to the extent allowed by antitrust authorities or the court enforcing the modified final judgment (MFJ).[7]

Cable-Telephone Company Competition

The only two telecommunication wires in most residences today are the telephone company's copper wires and the cable company's coaxial drop line. Each carrier is virtually ubiquitous, passing all but a few of the nation's homes. With costs of $1,000 or more per subscriber for switched-access telephone systems (much of which is an irretrievable, sunk cost) and the problems of arranging for right-of-way or conduit space, a third or fourth wire-based carrier will probably not develop outside of an urban area's principal business corridors. As a result, competition among wire-based carriers in residential and small commercial telecommunications markets may be limited to these two players for the foreseeable future. As new digital cellular systems, personal communications networks (PCNs), or even satellite-based wireless systems are built, competition for narrowband services may expand beyond these wire-

6. The regional Bell operating companies (RBOCs) were granted relief in 1990 (*U.S.* v. *Western Electric, et al., and Pacific Telesis Group, et al.*, U.S. Court of Appeals for the D.C. Circuit, April 3, 1990) from the line-of-business restrictions on information services imposed by the 1982 modified final judgment (MFJ). Since that time, they have not been particularly successful in developing new electronic Yellow Pages, want ads, or other electronic databases for general use.

7. This is precisely the strategy followed by many of the RBOCs in negotiating joint-venture arrangements with cable companies to exploit the possibilities of broadband communications outside their franchise areas. But they have not proceeded in a similar fashion with their own telephone networks, in no small part because of the 1984 Cable Act's restrictions on cable television for in-region operations.

based carriers, just as direct broadcast satellite (DBS) may provide the competition for wideband services.

At first glance, competition between the two wire-based technologies appears unlikely. The cable television plant is a tree-branch, one-way plant with limited return bandwidth and no switches; the telephone plant is a two-way network with limited bandwidth. The difference in management orientation is perhaps equally important. Cable companies are entrepreneurial enterprises, marketing an increasing variety of entertainment, information, shopping, and civic programming, but with relatively small technical and engineering operations. The telephone companies are regulated, largely nonentrepreneurial companies, with large technical engineering departments but relatively limited marketing skills.

These differences are now beginning to narrow, aided in part by joint ventures among the two groups. U.S. and Canadian telephone companies are moving outside their telephone franchise areas to enter joint ventures with cable companies that will convert the cable television plant into a full-service information network, offering some combination of switched voice-data and video programming.[8] At the same time, many large telephone companies are beginning to convert their franchised telephone operations to a mixed narrowband/broadband service. This new service could offer video-on-demand, as allowed on an experimental basis in the United States by the FCC.[9]

The convergence in technology is further facilitating competition between these two industries. Telephone companies are rapidly converting their entire backbone transmission plant to fiber optics. In addition, as copper wire degrades in their feeder plant, they are replacing it with fiber, thereby extending fiber optics closer to the consumer.[10] Similarly, cable companies are replacing

8. Most of the RBOCs and Bell Canada Enterprises have entered into joint ventures with one or more cable companies (see chapter 1).

9. Although the FCC has approved several such proposed systems, the telephone companies have begun to postpone their plans to build fiber or fiber/coax interactive video networks as this book goes to press.

10. The FCC's *Statistics of Communications Common Carriers*, 1994–95 edition, reports that telephone carriers had extended nearly 400,000 sheath kilometers of fiber by the end of 1992, containing 16.1 million km of fiber.

their distribution plant with fiber optics. Both entities are thus moving toward having fiber approach the final distribution node. As they extend fiber in this fashion, both have opportunities to convert their plant to an architecture that allows them to compete.

Telephone Networks

Telephone companies can essentially choose one of three alternatives to begin offering broadband services. They could simply string coaxial cable from a node (the curb) that is close to the subscriber's residence or business, overlaying the existing paired-wire drop line but not replacing it. They could upgrade the copper distribution loop to permit delivery of one-way video, perhaps through ADSL.[11] Or they could simply replace the rest of their copper plant with fiber optics and perhaps coaxial cable for the final subscriber drop line, offering an integrated voice, data, and video service over the same line.

The last of these three telephone company strategies is creating new regulatory nightmares. Pacific Bell has announced plans to rebuild its California plant for $16 billion, installing an integrated fiber-optics/coaxial-cable system to the final subscriber.[12] The cost of this plan is estimated at less than $1,000 per subscriber, but participants in the related California and FCC regulatory proceedings have already begun debating how this cost should be "allocated" among services.[13] The cable companies obviously want as large a share as possible assigned to video services, but Pacific Bell has offered to allocate only $137 per line to video dial tone. As in other similar examples of monopoly-competition interface, regulators will be forced to choose among structural separations, rate caps, or other safeguards to settle this dispute.

There is perhaps no better example of the inability to learn from history than the current disputes over the proper apportionment of fixed and common costs. We appear to be preparing to replay the

11. ADSL has been rejected by Ameritech as too expensive (*Telephony*, February 21, 1994, p. 17), but recent developments may reduce this cost from $2,000 per line to $500 per line by 1997 ("Around the Loop," *Telephony*, August 28, 1995, p. 6).

12. See FCC, In re Applications of Pacific Bell . . . , File Nos. W-P-C 6929, January 31, 1994. However, Pacific Bell's commitment to this project is now in doubt.

13. See Richard Karpinski, "The Video Dial Tone Firestorm," *Telephony*, February 21, 1994, pp. 28–30.

futile search in the 1960s and 1970s for methods to apportion AT&T's fixed and common costs, an exercise that ended in failure. It is essential that alternative regulatory schemes be developed to avoid this fruitless exercise. In this section we examine some of these, such as open network access (ONA) and unbundling.

Cable Networks

Cable companies are also poised to adapt their networks to provide a full array of telecommunications services. By building fiber optics into their feeder plant, they can establish separate distribution nodes at various distances from their headends. Each service area would have sufficient bandwidth to offer both one-way cable television services and two-way switched voice-data services. By installing switches at the various nodes, cable companies would become telephone companies as well. The final distribution of the telephone signal could be handled over existing coaxial cable or through a new spectrum-based PCN system.

For cable companies, whose one-way cable television rates are now regulated, a number of new regulatory hurdles may arise. If they are to offer switched telephony services, their rates and public-service obligations may be controlled by state regulators.[14] Second, under the terms of the 1984 Cable Act cable companies may be subject to a telephone company–cable cross ownership ban in the same franchise area. All of the issues involving cross subsidies could also arise when these companies are subjected to cost-of-service rate reviews under the 1992 Cable Act.

The likely success of telephone services offered by cable companies can be gauged by examining the experience of cable television companies in the United Kingdom. Cable television was launched there *after* liberalization of telephony; therefore, cable companies could begin offering telephony service from the beginning. The result has been startling. By pricing telephony services at 10 to 15 percent less than rates charged by British Telecom and by arranging long-distance interconnection with Mercury, U.K. cable companies have been able to enroll about 20 percent of their

14. At the very least, they will have to file tariffs with state and federal regulators for telephone services.

cable subscribers in their local telephone service. This share is expected to increase with time.[15] Their success is partly because they have no universal service obligation; nor do they have to pay access charges to British Telecom that include a contribution charge. Cable companies in the United Kingdom have been so successful that British Telecom has reconsidered its pricing policies to slow the pace of customer defections.

Legal Impediments

The 1982 MFJ that settled the AT&T antitrust suit has limited the geographical domain of the divested RBOCs for more than 10 years. These companies are forbidden to offer services to points outside their local access and transport areas (LATAs), including all forms of video, voice, and data services that cross LATA boundaries. In practice, this means that the RBOCs cannot operate in video markets without a waiver from the decree's restrictions, even though such participation may be entirely outside their telephone franchise areas. Video programming is delivered to terrestrial distribution points by satellite, whose national footprints obviously cross virtually all LATA boundaries. Moreover, the RBOC investment in cable companies violates the inter-LATA ban, because cable-system boundaries are not drawn to conform with the LATAs devised in the 1982–84 period.

The MFJ provides two other possible limitations to RBOC participation in broadband competition. First, the inter-LATA restriction prevents any RBOC venture from transmitting information services over long distances from a central database. If these information services are to be provided by the carrier, not by a third party that leases space on it, RBOCs would be prohibited from such a venture without a waiver. The new interactive information networks also require a variety of new equipment (such as servers and consumer terminals) whose design may be intrinsic to the success of the new multipurpose network. The manufacturing-line-of-business restriction literally prohibits RBOCs from par-

15. See the affidavit of John Anderson Kay submitted with the Motion of Bell Atlantic, Bell South Corporation, NYNEX Corporation, and Southwestern Bell to Vacate the Decree, *U.S.* v. *Western Electric and American Telephone & Telegraph Company*, U.S. District Court for the District of Columbia, Civil Action No. 82-0192, July 6, 1994.

ticipating in the manufacture or even the development of such equipment.

Finally, as we have noted, the 1984 Cable Act provides a legal impediment to competition by banning U.S. telephone companies from controlling program choices on their networks. This ban has been successfully attacked in one appellate court, however; it may be repealed in forthcoming legislation.

The Monopoly Bottleneck

If integration of local access, long distance, and equipment production provided opportunities for AT&T to abuse its regulated-monopoly position, the expansion of current telephone companies into much more sophisticated services raises questions of equal importance. Is it desirable to let local telephone companies build subscriber links to the information superhighway? Should they be allowed to invest in the consumer and business services that are delivered over this system as well as provide access for others, including services directly competitive with their own? Similar issues exist where cable companies seek to become the "local loop" on the information superhighway.

There are a number of options available for controlling the apparent problems of exploitation of local-access bottlenecks as the telecommunications network evolves. But the first and perhaps most crucial question is whether the local-access portion of the network is likely to always be a natural monopoly, and if such monopoly is sufficient to justify regulation. If the local-access provider is indeed a natural monopoly, must it be structurally and institutionally separate from the rest of the network?

Natural Monopoly

As we have argued elsewhere in this book, it is difficult to know if the local loop constitutes a natural monopoly in the traditional sense (that is, whether local access service is most efficiently provided by one firm rather than by two or more rivals).[16] With resi-

16. There is one strand of the recent empirical literature that argues that tele-

dential access rates set below average incremental costs, market entry of firms with new access technologies is necessarily impeded. Moreover, state regulatory hostility toward competition in the local loop has been considerable, even in "deregulated" Nebraska.

Today, nearly two-thirds of all homes have two communications lines coming into their homes: the telephone drop line and a coaxial cable for cable television. If the local *loop* is the natural-monopoly bottleneck, this monopoly is already a duopoly. These wires connect to facilities less likely to have natural-monopoly characteristics—namely, telephone switches and cable headends. There are economies of scale in providing these distributive functions, but such economies are limited. Most metropolitan areas have multiple local switching centers and many cable headends. Competitive switching facilities already exist in the form of private branch exchanges (PBXs) and competitive-access provider (CAP) switches located close to the telephone company's switch. Moreover, local transport between switches, including local Class 5 switches and long-distance Class 4 switches, is becoming competitive.

Unfortunately, much of the discussion about competition or the contestability of the local loop misses an essential point. *The lack of multiple sellers of local access is not the root cause of the bottleneck problem.* Suppose for a moment that the municipal government owned the drop line and made it available to a variety of users at a nominal charge in the same way that it owns the transportation bottleneck, the local streets. Would access to the subscriber be made easier for competitors, just as competitive pizza delivery, package express, or plumbing and heating contractors have easy access to households and businesses by local roads? The answer is quite clear: the bottleneck is not the telephone drop line but the interface between this line and those wishing to transmit signals over it. Even if the municipality maintained a policy of allowing open access to these lines, it could not deliver on this policy unless

---

communications services, even in local access, are not subadditive. (See John S. Ying and Richard T. Shin, "Viable Competition in Local Telephony: Superadditive Costs in the Post-Divestiture Period," paper presented at ENSAE-CREST (International Conference on the Economics of Radio-Based Communications), Paris, France, June 1994.) This view, however, is a decidedly minority opinion among students of telecommunications.

Figure 8-1. Configuration of a Local Network in a Competitive Era

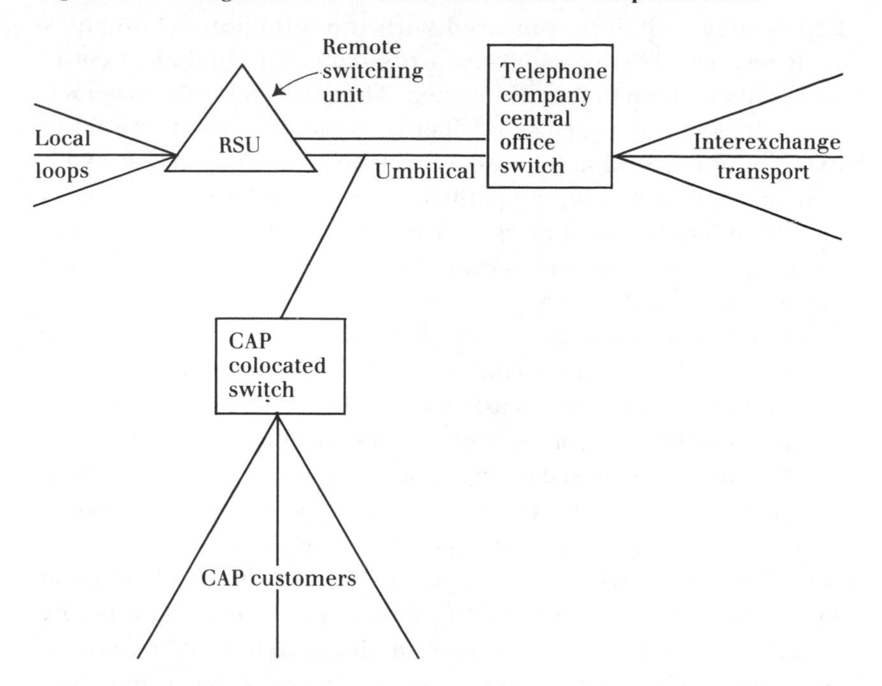

it also controlled the interface of the lines with the national telephone network. Access to the telecommunications network is much more complicated than access to the highway infrastructure (and becoming even more so). Highway access can be provided by driveways, alleys, or access ramps on private property; the local telephone company's bottleneck is the control over the interface, somewhere near the switch or at the switch itself. But it is not economies of scale in the switch per se, because switching may indeed be a potentially competitive activity with scale economies that are exhausted at a level far below that of the entire local market.[17]

The evolving competition in the telecommunications network is highlighted in figure 8-1. A subscriber is connected to the telephone

17. For an interesting discussion of this issue, see Economics and Technology, Inc./Hatfield Associates, *The Enduring Bottleneck: Monopoly Power and the Local Exchange Carriers*, study prepared for AT&T, MCI, and CompTel (Boston, February 1994), chapter 2. Note that the fact that competitive-access providers (CAPs) ask for co-location, connecting their own switches to those of the telephone company, suggests that switching is not a natural monopoly.

company's network through a remote switching unit (RSU) or a CAP whose switch is colocated with the telephone company's switches. The RSU concentrates calls along an "umbilical cord" that delivers them to the LEC switch. The CAP's network may look much the same, but each subscriber is connected to only one of the two competing local-access providers who have equal access to the national network. The competitive interexchange (IX) providers gain access to the local market through trunk-side connections to the LEC switch, from which they are connected to the subscribers of the LECs and the CAPs.

Even if competitive-access technologies develop and thrive, the bottleneck problem in telecommunications will not necessarily be solved. For example, let us assume that three or four competitors of equal size take half of each local market's telephone subscribers and that more are poised to enter the market. There are, however, a number of prices that must be controlled by competition or some other means. For instance, competition may put a ceiling on subscription rates to the network, but it may not be as effective in preventing carriers from exploiting their power over *terminating* calls. Each of the four or five local carriers can only offer communications services if the called party can be connected at the other end. But unless there are multiple connections to each subscriber, each company has a monopoly over the sale of these terminating services. It is this power—not the existence of increasing returns to scale—that may create the bottleneck. Competition from several LECs may not solve this "monopoly" problem.

If there are five competitive LECs (the incumbent plus four new entrants shown as CAPs), each will offer a menu of rates. But, as we have discussed, each LEC is likely to have monopoly power over the termination of incoming traffic to its subscribers. Except for 800 service or cellular service in North America,[18] incoming calls are free to the subscriber but are priced to the *caller*. Taylor quantifies several types of externalities in telecommunications networks.[19] The first is the "network externality"—that is, the addition of another subscriber in a small network increases the value of

18. Cellular service is priced differently in Europe. Incoming calls are not paid for by the recipient.

19. Lester D. Taylor, *Telecommunications Demand in Theory and Practice* (Amsterdam: Kluwer Academic Publishers, 1994).

access to existing or potential subscribers. The second is the "call externality"—a call to another person may generate a return call to the caller or to third parties. A third type of externality not analyzed in detail by Taylor is the decision to answer an incoming call of unknown origin. There are the potential benefits of hearing from a friend or, alternatively, the unexpected costs of hearing an unwelcome telemarketing solicitation. A subscriber may choose not to answer, to use an answering machine to screen calls, or to subscribe to a caller identification service in response to this unpriced externality of incoming calls. Moreover, costs are imposed on the calling party if no one answers; incoming calls thereby create externalities for both callers and their intended recipients.

Imagine a situation in which each of five competitive local-access providers is vying for subscribers. Each of these companies must arrange to terminate calls on each of the other's systems. There are several ways a market may be organized for this purpose. One is the current arrangement in long-distance services—the caller pays to terminate calls. In this type of market, a given LEC may demand high termination fees to put its rivals at a disadvantage, particularly if individual subscribers are unaware of the fees charged for terminating other carriers' calls. How do competitive LECs make the subscribers of a competitor aware of these high termination fees? Perhaps more important, will these subscribers care that their carrier is imposing large costs on the subscribers of other systems who want to call them? In short, the market for termination in this institutional arrangement will probably be highly imperfect.

But there is another approach to organizing this market that could easily be imposed by regulators. The price of termination could be paid by the *recipient* of those calls, not by the caller. Under this arrangement, a given LEC cannot "hold up"other access providers with high termination fees. With the subscriber paying for incoming calls, no carrier can exercise market power. Subscribers will be faced with price menus from all competitive carriers that include the rates for incoming calls.

Aside from North American cellular markets, there are few markets in which the receiving party pays the delivery charges for communications. Telephone calls, faxes, mail services, and telegrams are paid for in their entirety by the sender, even though the recipient benefits from the communication. These market solutions are likely to be inefficient. However, charging the recipient for incoming calls will also likely be inefficient, leading to greater use of display and other technologies to monitor these calls. However, the ability to ward off undesired incoming calls is not an inefficiency but a benefit that derives from charging the recipient for call termination. There are transaction costs and inefficiencies from incorrectly screening out desired calls that must be compared with the inefficiencies and monopoly distortions derived from charging only the originating party for all calls. Of course, if recipients were billed for termination of routine calls, businesses would ask for and possibly receive new "reverse 800" services, paying termination fees to gain access to subscribers.

At this point, the discussion may seem confusing or even amusing. However, the issue is important: communications involves the joint provision of services to caller and recipient under conditions of inadequate information. Alternative pricing schemes for these services can be developed. The current pricing scheme for telephone services, generally requiring callers to pay all variable charges for calls, is inefficient and perhaps unsustainable in the face of competition at the local exchange.

The principal competitive problem in this network industry is therefore the control over access that the company with the customer-line interface has. Even if every customer owned his or her own access line, someone would have to deliver signals from or to this line. Somehow, the customer must be interconnected to the switching hierarchy that guarantees ubiquitous access to other subscribers through outgoing or incoming calls. For competition to blossom in telecommunications, there must be some way of assuring entrants access to all subscribers with whom their own customers want to interact. This requires either that each competitor have its own direct access to all required subscribers or that equal access be obtainable through market transactions or guaranteed by regulation.

As the information superhighway grows, any natural monopoly that derives from control of paired copper-wire subscriber access lines may be superseded by superior forms of access. Higher-speed services may require fiber-optics or coaxial drop lines, or even an antenna to receive signals over the electromagnetic spectrum. As these higher-speed technologies develop, the optimal subscriber distribution technology may also have natural-monopoly characteristics, but it may be very different from current network technology. But for the moment, the most important telecommunications policy issue in the United States (and imminently in Canada) is the assurance of access to the switched national (narrowband) network.

Structural Solutions

There are two potential structural solutions to the telecommunications bottleneck problem: direct connections for all carriers, or structural separation of the bottleneck from other telecommunications activities. The direct-connection solution is commonly found in the transportation sector. Airlines build their own terminals at airports; delivery services own their own vehicles and distribution facilities. In each case, sharing the bottleneck might be more efficient, but duplication of facilities guarantees that each carrier can control its own traffic and service quality. Structural separation is also found in transportation markets. The railroad carriers typically have not owned passenger stations; local bus companies often do not own their terminal facilities. In both cases, public ownership of the bottleneck allows (but does not guarantee) equal access to competing carriers who need to interconnect.

Direct Access to Subscribers

Although it is possible that each local-access supplier could connect *directly* with its own subscribers (that is, at the subscribers' premises), such direct connections are unlikely for terminating their traffic at other subscribers' premises. A spectrum-based carrier could conceivably offer ubiquitous connections to all households and businesses. But even a system like Motorola's proposed IRIDIUM Project will probably not enroll every telephone subscriber in the country. It will still need to connect with other *carriers* to terminate its own traffic as well as traffic that originates from other carriers' subscribers.

One possibility is for each subscriber to own a local distribution loop. But as we have seen, this would require consumer ownership of the network interface (that is, the first switching node) as well. In fact, because local distribution lines may be a complicated combination of various cables that carry several hundred (or even thousand) subscribers' lines, no single subscriber could own and maintain his or her own line except in cooperation with all other nearby subscribers. Customer ownership would therefore require a telephone cooperative or some other device. This is similar in spirit to municipal ownership of the roads connected to state and interstate highways.

Structural Separation

The solution to the bottleneck-monopoly problem used to settle the AT&T case was structural separation of the bottleneck-monopoly function from the competitive activities of AT&T. The BOCs were separated from the long-distance operations of AT&T, but a large amount of potentially competitive local (Class 5 or end-office) switching and transport was left within the BOCs. A division that left less activity with the separated natural-monopoly carriers would have been more consistent with the goal of isolating competitive and monopoly functions.

The problem with any structural separation rule is that it leaves the monopolist with a highly circumscribed set of functions and a limiting plant investment. Such separation would presumably be continued until there was competition in the local loop, however defined. Under this constraint, the local monopolist would have little incentive to upgrade its plant to make it capable of delivering new or higher-speed services, because the monopolist could not profit from any of the new vertically related services. Indeed, regulators would be skeptical that bold steps to embrace new technology would lead to a larger "revenue requirement," without any assurances that a market to cover these new revenue needs would in fact exist. Ironically, it might be in the bottleneck carrier's interest to forgo new investment and be declassified as a monopoly bottleneck when an entrant moves in to offer the new service.

The most aggressive of regulatory policies to isolate the local loop from the rest of the network would also compel the regulator to price access to cover its full costs. Once the local loop is isolated

Table 8-1. Real Investment by Local-Exchange Companies in the United States, 1984–94

Year	*Investment (billions of 1987 dollars)*	*Share of total telephone investment*
1984	16.0	.70
1985	20.0	.78
1986	19.3	.70
1987	18.5	.70
1988	17.5	.72
1989	16.1	.70
1990	17.4	.77
1991	17.4	.81
1992	17.4	.76
1993	17.0	.75
1994	16.8	.60

Sources: FCC, *Statistics of Communications Common Carriers*; Department of Commerce, Bureau of Economic Analysis.

and all other services are deregulated, cross subsidies become more difficult, if not impossible.[20]

The MFJ's Structural Separations

The line-of-business restrictions in the AT&T divestiture decree were even less severe than those described here for isolating only the presumed local-loop monopoly. The divested BOCs have been able to offer a variety of central office services (such call waiting, call forwarding, and conferencing) as well as an array of intra-LATA long-distance services. They kept their cellular, terminal-equipment sales, and rental businesses, and even their Yellow Pages, and they have even been freed to develop information services. All of these activities are related in some way to their local access-exchange plant, so they have not had the incentive to neglect investment in their local plant. Nevertheless, local-exchange investment has clearly stagnated since the 1984 divestiture, although not as a share of total telecommunications investment until recently (table 8-1).

Nor have the BOCs been able to plead poverty as a result of the limitation imposed by the 1984 decree. They have remained profitable and have outperformed the market on a risk-adjusted basis

20. This was the goal of the original Rochester Telephone proposal.

Table 8-2. Average Annual Returns from Holding Common Equities of the Regional Bell Operating Companies (RBOCs), 1984–93

Company	*1984 (third quarter)–93*	*1989–93*
Ameritech	.220	.177
Bell Atlantic	.218	.184
Bell South	.196	.152
NYNEX	.192	.117
Pacific Telesis	.239	.195
Southwestern Bell	.246	.235
U S West	.210	.173
Estimated beta[a]	.68	.78
Risk-adjusted average[b]	.319	.226
Average for all stocks	.167	.154

Sources: Center for Research in Security Prices, Chicago; R. G. Ibbotson Associates, *Stocks, Bonds, Bills, and Inflation, 1993 Yearbook.*
a. Estimated for equal-weighted portfolio of all seven RBOCs.
b. Equal-weighted average return divided by estimated beta.

since 1984 (table 8-2). Their pleas for relaxation of the decree's line-of-business restrictions have been largely based on the potential positive effects of such a relaxation on long-distance communications and other services.[21]

Other regulatory jurisdictions have *not* imposed structural separation to deal with the onrush of competition. In the United Kingdom, OFTEL, the regulator of British Telecom, has stated: "BT [British Telecom] has about 97 percent of local terminations and the most fully developed long-distance network in the U.K. . . . Structural separation would not solve all of the problems of interconnection by itself. Moreover, there may well be benefits to telecommunications users from the exploitation of economies of scope by companies offering integrated network and retail services. *OFTEL therefore does not advocate structural separations*" [emphasis added].[22]

In Canada, neither the regulator nor the major competitor to the land-line telephone companies has argued for structural separation. No other country in Europe or Asia (including New Zealand and Australia) has opted for structural separation as the means of regulating nondiscriminatory access to bottleneck facilities.

21. Their requests for relief from the manufacturing restriction are based largely on their asserted need to develop equipment to deliver new video and information services.

22. OFTEL, *Interconnection and Accounting Separation.* Consultative document issued by the Director General of Communications (London, June 1993), pp. 2, 5.

Unbundling and Mandatory Interconnection

Even if the local loop eventually proves to be a contestable market (in that several sellers could compete for or in the market successfully), the nature of telecommunications as a network industry requires some form of interconnection. Our central argument is that no customer wants to be connected to a local carrier that cannot interconnect with other carriers in his or her own geographic area or elsewhere. The decree separating the BOCs from AT&T has a provision requiring divested BOCs to implement equal access for all *interexchange* carriers, and the FCC subsequently imposed such a requirement on all local-exchange companies.[23] In a competitive world, such a regulation may still be necessary. But how are regulators to determine what equal access is and the reasonable rates for providing it?

Given the complicated nature of modern telecommunications, enforcing mandatory access on equal terms to *every* carrier (including the access provider's affiliates) may prove extremely controversial. The fear that access providers will evade any such rule has led to the continued separation of the BOC facilities from those of long-distance carriers. If the LECs are integrated into long-distance services and other vertically related services, they could possibly develop and operate their networks to favor their own vertically related operations. However, GTE's recent experience with vertical integration was remarkably uncontroversial.

GTE is one of the largest LECs in North America. It has owned substantial equipment manufacturing facilities and, until recently, a large share of Sprint, the third largest U.S. long-distance carrier.[24] After 1984, GTE was forced to provide equal access in most of its switching centers, with quite favorable results. There were apparently no serious complaints that GTE had favored Sprint in providing access to interexchange carriers thereafter. Moreover, most large LECs offer their own cellular service and provide access to their wire-line networks for their direct competitors, the second cellular companies in each market. Despite obvious incentives to

23. Before the FCC, Washington, FCC 92-440, CC Docket no. 91-141, In the Matter of Expanded Interconnection with Local Telephone Company Facilities.

24. Much of the manufacturing operation has been sold, and GTE slowly sold its interest in Sprint to its partner, United Telecom. GTE now has no interest in Sprint.

favor their own cellular operations, the LECs apparently do not. The independent cellular companies currently have more than 52 percent of subscribers (non-wire-line A frequencies) in the top ten markets in the country.[25]

The modern LEC offers a myriad of services through its network. These include connecting calls within its own franchise area, access services to long-distance carriers, various call-management services (such as call waiting and call forwarding), billing services to long-distance carriers, and a number of dedicated-line services to business customers. In addition, most LECs are developing information services, such as electronic Yellow Pages, or want ads, and access to shopping and news services. In delivering these services, the carriers incur large joint and common costs, making it difficult to estimate the incremental costs of connecting other carriers to a network. Any carrier access rate established by this local company can therefore always be criticized as unfair, particularly if the local telephone company also competes with the carriers it is offering to connect to its local network.

One way to minimize these disputes is to separate the access operations from all others in some way. But, as OFTEL has correctly stated, even structural separations do not solve all problems of interconnection. Even if the local-access company provides *nothing* else, the pricing issues remain. Providing nothing but access is not a feasible alternative, particularly in an era of such rapid technical change in telecommunications. Solutions to the pricing issues must be found.

Unbundling

Under the FCC's computer rules, telephone carriers no longer have to maintain structural separations of their monopoly and competitive activities. They must instead unbundle and offer open network architecture (ONA). The unbundling of the network elements is largely left to the carrier, and there is considerable controversy over the selection of the unbundled basic service elements (BSEs) in the carrier's offering. More important, as services are divided

25. Herschel Shosteck Associates, Ltd., "Data Flash," *Cellular Market Quarterly Review*, vol. 8, no. 2 (December 1994), figure 3.1.

into ever narrower subcategories, no regulator can hope to precisely determine the reasonableness of the charges for each element. Nevertheless, ONA marks a continuation of the trend toward introducing competition into the network and isolating the monopoly bottleneck. With unbundling, the regulated telephone companies are allowing competitors to pick and choose among services they can provide and those still supplied most efficiently by the franchised telephone monopolist.

In the past two years, two companies have proposed a substantial unbundling of their networks to obtain some regulatory relief. In one case, Rochester Telephone (now Frontier) proposed isolating its local telephone services in a structurally separate subsidiary that offers virtually no retail services.[26] The retail services would be offered by a lightly regulated subsidiary company that offers competitive services, including retail local service. In the other case, Ameritech proposed unbundling its local-service offerings to be permitted to offer interexchange services across its entire five-state region.[27]

Neither offer was willingly accepted by regulators. The New York State Public Service Commission finally approved a substantially modified version of Rochester Telephone's proposal in 1994.[28] As this book goes to press, the U.S. Department of Justice is finally supporting a restricted "experimental" version of the Ameritech plan, limited to a small geographic area and allowing Ameritech to only offer resale of inter-LATA services.[29]

Many published discussions and analyses have examined the degree to which unbundling and open network architecture can solve the problems of market power in a vertically integrated telecommunications sector.[30] These ideas are being discussed in the United States, Canada, the United Kingdom, New Zealand, and Australia. The issues are complex. One example is setting a price on the components of unbundled access. In New Zealand, a lower

26. State of New York Public Service Commission, "Petition of Rochester Telephone Corporation for a Proposed Restructuring Plan," February 3, 1993. For further details, see chapter 2.

27. See FCC, Petition for Declaratory Ruling and Related Waivers to Establish a New Regulatory Model for the Ameritech Region, March 1, 1993.

28. *Joint Stipulation and Agreement*, April 1994.

29. Once again, more details are in chapter 2, above.

30. See, for example, Aspen Institute, *Local Competition: Options for Reform*, Forum Report, 1994.

court accepted the Baumol-Willig criterion for pricing a bottleneck service in the presence of cross subsidies. This "efficient-component" rule allows the regulated carrier to price certain service elements (such as terminating access) at rates that include a contribution or subsidy charge, as long as the regulated firm uses this same rate in pricing its own retail services that use the same service elements.[31] However, this approach was subsequently rejected by the Court of Appeals but appealed to New Zealand's Supreme Court—the United Kingdom's Privy Council, which upheld the lower court. A similar proposal was part of Rochester Telephone's revised proposal that was ultimately accepted by the New York State Public Service Commission (NYSPSC) in 1994.

Additional unresolved issues include the degree of required unbundling, the points at which competitors interconnect, and the determination of incremental costs of unbundled elements. With unbundling, it becomes more difficult to maintain cross subsidies to serve purported social goals. Attempting to measure the incremental cost of unbundled service elements is obviously difficult, if not impossible. As we have pointed out, regulators were not successful in determining the costs of *services* in the 1960s and 1970s; they are no more likely to be able to estimate the costs of the rapidly changing service elements that make up the exploding array of service offerings from this sector.

Compulsory Interconnection

Regardless of the prospective rules for regulating or deregulating incumbent carriers, the telecommunications network is likely to remain fragmented and to become even more so. Competitive long-distance carriers, specialized value-added (information services) networks, LANs, and competitive local carriers require interconnection arrangements to function efficiently. As the local network fragments, these requirements intensify, as the FCC's colocation proceeding demonstrates.

Mandatory interconnection is necessarily a regulatory exercise. The mandate means nothing if it is not required at a "reasonable"

31. See William J. Baumol and J. Gregory Sidak, *Toward Competition in Local Telephony* (MIT Press and American Enterprise Institute, 1994), chapter 6.

rate. Regulators must therefore specify not only the nature of interconnection required but also the terms under which it is to be made. The regulation of these interconnection rates is no simple task, as we have pointed out, and it provides ample opportunities for politically driven distortions. The best examples of such distortions are intrastate access charges for competitive long-distance carriers. In 1993, the common-line components of these charges varied from as little as 1 cent per minute to more than 6 cents per minute.[32] Surely the "cost" of using a fixed subscriber-access line did not vary by a factor of six.[33]

It is possible that there will eventually be so much competition at all levels of telecommunications that mandatory interconnection will be unnecessary. Under such conditions, carriers would be compelled by commercial realities to interconnect with one another, particularly if subscribers were billed for incoming calls. At present, however, it is difficult to avoid the necessity of regulated, mandated interconnection.

None of these policies is perfect. Complete separation sacrifices economies of scale and scope in delivering telecommunications services. Accounting separations may simply enmesh the regulator in constant disputes. The choice of policy is necessarily a compromise among competing objectives—ensuring that new competitors with new technologies are not discouraged from entering the market, and yet allowing incumbent local-access providers to exploit economies of offering similar services over their own networks. At the same time, policymakers should factor in the prospective benefits of allowing markets to correct the distorted rate structures that pervade this sector after decades of regulation.

Effects of Alternative (De)Regulatory Approaches

The four following strategies for dealing with the continuing problem of bottleneck monopoly succinctly embrace the range of choices open to policymakers:

32. Data provided by a large telecommunications firm and used in the chapter 5 empirical analysis.

33. The cost of the subscriber loop is entirely non-traffic-sensitive. The costs of switching and transport, on the other hand, do vary with use.

—Total deregulation without any mandatory access.

—A structural separation along very narrow boundaries, requiring that access providers only offer the lowest degree of switching. Carriers would provide open-network access to all competitors at regulated rates in an unbundled format.

—Structural separation within a holding-company framework (such as that advanced by Rochester Telephone), with total deregulation of all services other than the unbundled wholesale services.

—Accounting separations between monopoly and competitive (or wholesale and retail) services, combined with rate caps, compulsory access, and efficient components pricing to prevent predation and discrimination. (This is the approach used by OFTEL and proposed for Canada, Rochester Telephone, and other U.S. jurisdictions.)

TOTAL DEREGULATION AND MANDATORY ACCESS. Total deregulation does not seem practicable because of political and regulatory constraints. In both the United States and Canada, regulators (including Judge Harold Greene) will probably be unwilling or unable to deregulate all telecommunications. Nor is it clear that immediate deregulation would necessarily enhance economic welfare. Entry into local markets by new wire-based carriers may well be frustrated by the incumbent local telephone and cable companies' high ratio of sunk-to-total costs. As a result, local rates would likely rise substantially, disciplined only by a fear of reregulation or entry (and expansion) of wireless carriers, such as cellular and PCS services.

Equally important is the power incumbents will have to discriminate in providing termination services unless recipients are required to pay for incoming calls. If mandatory access is defined to include the requirement that recipients pay the termination costs for incoming calls, this exertion of monopoly power will probably wither more quickly.

It must be pointed out that local rates could rise significantly without reducing economic welfare if the consequence of deregulation were to reduce long-distance rates substantially. Without a universal-service obligation, new access providers would surely compete for originating and terminating the traffic of large long-distance users, depressing these rates greatly. Without contribution payments (the non-traffic-sensitive component of access charges), the incremental cost of long-distance service is probably no more

Table 8-3. Alternative Scenarios for Total Deregulation of the U.S. Telephone Industry, 1993

Category	*Before deregulation*	*After deregulation*	
		Scenario I[a]	*Scenario II*[a]
Interstate toll			
Average rate[b]	0.157	0.100	0.050
Total minutes[c]	211	296	497
LEC minutes[c]	0	30	60
Access charge[b]	0.063	0.02	0.02
Intrastate toll			
Average rate[b]	0.171	0.10	0.05
Total minutes[c]	162	242	407
LEC minutes[c]	81	40	80
Access charge[b]	0.08	0.02	0.02
Local service			
Average residential rate	$19	$57	$40
Average business rate[d]	$42.50	$85	$50
Residential lines[e]	95.6	92.5	93.5
Business lines[e]	42.4	41.0	41.2
Change in consumer surplus[f]	—	−35.37	+44.71
Change in producer surplus[f]	—	+39.77	−14.80
Net welfare gain[f]	—	+4.40	+29.91

Source: Author's calculations.

a. The incremental cost of access is assumed to be $0.02 per minute in both scenarios. The incremental cost of long-distance service is assumed to be $0.08 per minute in scenario I and $0.03 per minute in scenario II (plus access charges). The incremental cost of local residential and business access is assumed to be $360 per year in both scenarios.

b. Dollars per minute.

c. Billions per year.

d. Dollars per month.

e. Millions.

f. Billions of dollars per year.

than 5 cents and surely no more than 10 cents per minute.[34] However, average U.S. rates are still about 16 cents per minute for interstate carriers, due in part to access rates that exceed incremental costs. In addition, intrastate rates are about 17 cents per minute, though the costs of intrastate service are slightly lower than the costs of interstate service.

In table 8-3, we offer two possible scenarios of total deregulation.[35] Scenario I is the more pessimistic. It assumes that after deregulation the average flat-rate residential rate would treble, and the average business rate would double, because new wireless

34. In 1993, AT&T reported marketing, customer service, and G&A costs almost as great as access costs, which in turn were between 6.5 and 7 cents per minute. A substantial share of the marketing and customer service costs per minute are surely endogenous, as we argue in chapter 5.

35. The elasticity and cost assumptions are described below.

technologies have substantially higher costs than do the current LECs. (These increases are many times greater than the increases witnessed in Nebraska after its "deregulation" of telephone service.) In this scenario, the price of long-distance service falls to $0.10 per minute and long-distance access rates to $0.02 per minute.

The second scenario in this table assumes that residential flat rates rise by more than 100 percent and business rates by only 18 percent. In both cases, the rates are above average incremental cost (assumed to be $30 per month). But in the second scenario, long-distance rates fall to $0.05 per minute as a result of entry from the LECs and other companies with lower access rates.

We calculate the changes in consumers' and producers' surplus under 1993 conditions. We assume that the price elasticity of long-distance demand is −0.75, the price elasticity of residential-access demand is −0.03, and the price elasticity of business-access demand is −0.01. There are substantial net welfare gains in both cases, but they are distributed quite differently in the two scenarios. Obviously, if local rates double or triple (scenario I), the LECs' producers' surplus rises sharply and the consumers' surplus falls sharply. Though the LECs' gains outweigh the losses of the long-distance carriers and consumers, we would not expect this deregulatory outcome to be politically appealing.

In the second scenario, the increase in consumers' surplus is enormous—nearly $45 billion per year. However, the loss in surplus registered by the long-distance companies (more than $18 billion)[36] would make this outcome hard to sell, even though the net improvement in economic welfare is almost $30 billion annually.[37] These two scenarios highlight the problem created by decades of misguided regulation: any movement toward efficient prices is likely to result in at least large transitional losses to some

36. This loss of surplus would undoubtedly be reflected in sharp declines in advertising outlays, G&A expenses, and even profit margins for long-distance carriers. We ignore cross-elasticities of demand among services and network externalities in these calculations.

37. This is much more than the net benefits of airline or surface-freight deregulation. See Clifford Winston, "Economic Deregulation: Days of Reckoning for Microeconomists," *Journal of Economic Literature*, vol. 31 (September 1993), pp. 1263–89.

market participants, even though the net impact on the economy would be favorable.[38]

This analysis assumes that "deregulation" of all services leads to sufficient entry into long-distance services and local access for long-distance users to lower long-distance rates to incremental cost. This is a reasonable set of assumptions for large business users but perhaps less so for residential services. However, with personal communications systems (PCS) looming ominously on the horizon, it would not be rational for the incumbent LECs to raise residential rates as high as we have assumed in scenario I of table 8-3. Given that at least one digital cellular company is already offering 1,200 minutes of peak service for $44 per month,[39] we would not expect rates of $48 per month to last very long, even if regulation disappeared and legislators ignored the trebling of these rates. However, we do not doubt that some light regulation of local services may be required—interconnection rules, requirements that recipients pay for terminations or that access charges fall by the amount of reduced universal-service contribution—to assure that these subscribers receive the benefits of increased long-distance competition.

NARROW STRUCTURAL SEPARATION. Some have suggested that the problems of inducing competition at the local-exchange level are so profound that market power will remain an issue in offering service to dispersed residential customers and access to long-distance providers.[40] Given this gloomy assessment, one possible solution would require local access to be totally separated from all other services. This, after all, was the implicit judgment that led to the AT&T divestiture, even though its execution was much less clear cut. In this solution, a local carrier would be left only with local loops, multiplexing, and the lowest degree of switching. We call this an NSSF (narrow structurally separated firm).

This potential solution is compelling in its simplicity but proba-

38. These calculations of potential welfare gains from deregulation and rate rebalancing are much greater than those in chapter 3, because the latter included only *intrastate* services. Most of the welfare gains come from repricing long-distance services, and the estimates in this chapter include gains from *interstate* long-distance services.

39. See the citation of Cybertel's tariff in chapter 7.

40. ETI/Hatfield Associates, *The Enduring Local Bottleneck*.

bly not the best choice, for at least two reasons. Any economies of scope between access and other services would be sacrificed. These economies could be great, particularly if the intelligence in the network is optimally located near the subscriber. Structural separation would ensure that this intelligence would have to be located elsewhere. It is perhaps for this reason (as we have stressed) that no nation has followed the lead of the United States in formally divesting other service operations from the LEC. In Canada, neither the regulator, the entrants, nor the Restrictive Trade Practices Commission[41] has recommended structural separation.

A second argument against the structural-separation solution is the effect that such separation would have on incentives and modernization. With its narrowly defined business, no opportunity for growth, and regulated rates, an NSSF would not be the most attractive company to manage. Nor would it be attractive to investors. They would see that entry and new technology would inevitably lead to a decline of the NSSF, because it could not embrace new market opportunities. It is difficult to imagine how such an institution would be able to adapt to the new technologies of the information superhighway. It could not possibly undertake the risk of building ADSL circuits at $2,000 each or extending coaxial cable to consumer interface units that must be designed for specific applications that the company cannot provide. The NSSF would inevitably be a stagnant, narrowband network, waiting to be superseded by the victor in the race for the information superhighway.

STRUCTURAL SEPARATION WITH A HOLDING COMPANY. A more attractive possibility is that the narrowly defined local network be established as a separate subsidiary of a holding company, offering wholesale local services to new entrants and the company's own competitive subsidiary. (This is precisely the proposal advanced by Rochester Telephone, but substantially modified in negotiations with the NYSPSC staff.)[42] Under this alternative, the wholesale bottleneck monopolist is a separate *subsidiary*, offering all unbundled elements of the local network under tariff. The competitive subsidiary of the company must use these tariffed rates in calculating its

41. Now the Competition Tribunal.

42. See the Petition of RTC before the New York State Public Service Commission, above.

own rates, while ensuring that its service rates cover these tariffs plus its own incremental costs.

Even under this proposal, there is a considerable risk of loss of economies of scope. If the regulated bottleneck is to be truly independent, it cannot join with the competitive subsidiary to plan facilities to exploit potential joint economies. If such coordination takes place, the regulated entity's investment decisions may be designed to favor its related competitive subsidiary.

ACCOUNTING SEPARATIONS. The United Kingdom and Australia are pursuing a less formal separation of wholesale and retail functions based simply on accounting and "efficient-component" pricing.

In the United Kingdom, OFTEL has enunciated four principles that could achieve "fair" competition but not necessarily efficient pricing:

—transparency of charges;

—sustainable charges that include an access-deficit contribution (ADC).

—nondiscrimination (that is, British Telecom must charge itself the same charges it levies on competitors); and

—equitable arrangements among competitors.

OFTEL has decided to separate wholesale and retail charges for British Telecom (BT). It requires unbundling for BT services and uses price caps for regulating BT's rates. All rates must be published, including the access charge and what BT charges its own retail operations. However, OFTEL has explicitly rejected "a purely cost-based approach to calculating network charges. . . . BT's retail charges are not at present aligned directly with costs."[43]

In Australia, the regulator has also opted for accounting separations, unbundling, and rules for nondiscrimination. Unlike the United Kingdom, however, Australia requires that the initial price of access for competitive carriers be the incumbent firm's long-run incremental costs. Once the entrant's market share reaches 25 percent, the price of access is negotiated between the new carrier and the incumbent. The cost of universal access is borne through a separate charge on all service providers.

AUSTEL, the regulator, has proposed revisions to its regulatory

43. OFTEL, June 1993, p. 8.

policies.[44] Three basic carrier services will be subject to rules on dominance: basic customer connection, facilities between the basic customer connection and the first point of interconnection; switching; and interexchange transport. All other services will not need to be tariffed because they are deemed to be competitive. Rates for the basic services (called flag prices) are to be set anywhere the carrier wants subject to the requirement that all of the incumbent's retail rates must reflect these charges.

This proposal generates a number of criticisms. As the carriers have pointed out, switching and interoffice transport are not obviously monopoly services. As we have indicated, it is also not obvious that termination involves market power if the charges are borne by the recipient. However, as we have emphasized, the externality in incoming calls makes any termination-charge system inefficient. Finally, allowing the carrier to price access at any level as long as the component-pricing rule is followed may lead to substantial bypass by competitive carriers and pressure to rebalance rates—a consequence perhaps not anticipated by AUSTEL.[45]

Concluding Policy Recommendations

Obviously there are no easy choices in this period of rapid technological change and evolution toward competition. But we can offer a number of unequivocal recommendations. For example, it is clear that the pressure for local competition and unbundling will require a fundamental restructuring of telephone rates. As networks are unbundled and entry occurs, bypass pressures will force a rebalancing of rates. Subsidies will have to be eliminated or more carefully targeted, and rates for retail services will have to move toward cost. The pressure from local competition and multiple entrants will also focus attention on the issue of pricing call termination. Access charges may have to be levied on the recipient, and these charges will be forced toward cost.

44. AUSTEL, Decision Making Framework, briefing paper for public meeting, June 20, 1994.

45. There are also other problems with the proposals such as what the extent of required unbundling would be, who will determine if requested further unbundling is feasible, and how rates will be set for this further unbundling.

None of these changes strikes us as pernicious or even politically dangerous once understood by the public. As we have shown here and in chapter 3, the welfare gains from rate rebalancing could be enormous. We should therefore welcome rate rebalancing and the new technology that competition will bring to the local market.

Our results show that the various attempts to introduce regulatory reform (such as rate caps or revenue sharing) have had little impact on the structure of rates, particularly in the U.S. intrastate jurisdiction. This lack of progress may be traced to the reluctance state regulators have toward allowing real competition in local-access and intra-LATA long-distance markets. The local telephone companies have accepted the burden of a woefully inefficient rate structure as part of a Faustian bargain in which they continue to receive protection from competition. But with wireless costs falling dramatically, this bargain is now crumbling. Competition is coming to the local loop, perhaps very soon. The local-exchange carriers and their regulators had better address the issue of rate distortions or face being overwhelmed by this new competition.

The FCC has achieved a modicum of rebalancing through the substitution of a subscriber line charge for part of the non-traffic-sensitive portion of the carrier-access rate. However, until recently state commissions have been reluctant even to allow full local access-exchange or intra-LATA competition, presumably fearing the effect on the rate structure. Canada's attempt to begin overhauling its woefully inefficient rate structure was temporarily blocked by its cabinet. Even now in Canada, local rates will rise by $2.00 per month on January 1, 1996, and another $2.00 per month on January 1, 1997, with a third phase (amount unknown) on January 1, 1998.[46]

Given the incredible complexity of current telecommunications technology, the rapid rate of change in this technology, and the proliferation of new services, we firmly believe that minimal rate regulation is required. In this era, monopolies are likely to be short lived, but regulation can impose substantial disincentives to invest and develop new products. With regulation comes the inevitable special pleading for protection from competition (as any observer of the intense Washington and Ottawa lobbying for changes in telecommunications legislation can attest). Markets are better

46. CRTC Telecom Decision 95-21.

judges of who has built the best mousetrap, not politicians and regulators.

Because telecommunications has strong networking properties, we are reluctant to recommend full structural separation as a preventive measure against vertical squeezes from incumbent monopolists. A regulatory regime should set well-defined ground rules for all current and prospective competitors, and competitors and entrants should be treated symmetrically. Our empirical analysis has shown that the current asymmetrical regulatory regime creates incredible inefficiencies, which both punish incumbents through onerous service requirements at noncompensatory rates and reward them through entry restrictions on more price-sensitive services. In addition, our analyses showed that halfhearted reforms such as partially flexible pricing do not necessarily move rates in the right direction. Nor does it benefit anyone other than the incumbent firm to partially deregulate without allowing full competition. Our statistical analyses of the intra-LATA market provided particularly strong evidence that flexibility without equal-access competition leads to higher and not lower rates.

In short, we support liberalized competition in all links of the information superhighway. The regulator should be left to specify interconnection rules, unbundling, an accounting separation between wholesale and retail functions for firms with bottleneck monopolies, and nondiscriminatory pricing for such integrated monopolists. With this minimalist regulatory framework in place, it should be left to the competition (that is, the antitrust) authorities in each country to arbitrate amidst the resulting cries of predation, discrimination in access, or unfair pricing. These authorities have more rigorous standards for determining whether a practice will reduce economic welfare than do regulators or legislators.

Index